TARTU SEMIOTICS LIBRARY 14

Tartu Semiootika Raamatukogu 14
Тартуская библиотека семиотики 14

Matemaatika kui modelleeriv süsteem: semiootiline vaade

Математика как моделирующая система: семиотический подход

MATHEMATICS AS A MODELING SYSTEM: A SEMIOTIC APPROACH

Marcel Danesi, *University of Toronto*

Mariana Bockarova, *Harvard University*

UNIVERSITY OF TARTU
PRESS
1632

Book series *Tartu Semiotics Library* editors:
Kalevi Kull, Silvi Salupere, Peeter Torop

Copy editor: Ene-Reet Soovik

Address of the editorial office:
Department of Semiotics
University of Tartu
Jakobi St. 2
Tartu 51014, Estonia

http://www.ut.ee/SOSE/tsl.html

ISSN 1406-4278
ISBN 978-9949-32-610-5

University of Tartu Press
www.tyk.ee

Contents

Preface

Mathematics and semiotics share many intellectual features and interests, from the study of how representations stand for specific kinds of referents to philosophical considerations of how these interrelate with reality. Nonetheless, rarely have in-depth studies of this intrinsic relation between the two been undertaken, with a few notable exceptions (as will be discussed in this book). Especially relevant to the study of the nature of mathematics is the concept of *model* – a term and notion that is used widely in both disciplines. However, to the best of our knowledge the theory of models in semiotics, known as *Modeling Systems Theory*, has rarely, if ever, been applied to the study of mathematical modeling. The purpose of this book is to do exactly that, since it is our view that mathematics is a *de facto* modeling system in the semiotic sense and it is our hope that from this it will be possible to gain considerable insights into how mathematics works and achieves the discoveries and forms of knowledge that it has since the dawn of antiquity. Hopefully, this will allow both mathematicians and semioticians to pursue similar or analogous research objectives with regard to understanding the biological and cognitive etiology sign systems and their connection to reality.

People have always taken pleasure in numbers and used numerical ideas to carry out the practical counting and arithmetical routines of everyday life. One of the oldest mathematical texts, an Egyptian work called the *Rhind Papyrus* written around 1650 BCE, is a collection of mathematical problems composed (seemingly) for educational purposes and to bring out the practical value of numeration and geometrical thinking to everyday life. The papyrus is also, indirectly, a treatise in how to do mathematics with the aid of symbols and notational devices. In a fundamental way, it is a treatise on the semiotic nature of mathematics, showing how symbolism (notation, numeral systems, and so on) is the sum and substance of mathematical method (Danesi 2002). In our view, any approach to understanding the nature of mathematics should take this very fact into account. Of course, the author of the papyrus never used any notions that could be construed as semiotic. But the work is inherently semiotic in the way it introduces mathematical concepts and problems. As this shows, it could be argued that from the advent of mathematics as a distinct discipline, mathematicians have been doing semiotics without knowing it. The late Thomas Sebeok (1920–2001) would often point out that the list of those who did semiotics without knowing it would fill the pages of countless books. He referred to this state of affairs as the "Monsieur Jourdain syndrome". Monsieur Jourdain was a character in Molière's *Bourgeois Gentilhomme* (1670) who, when told that he spoke good prose, replied that he was not aware that he was using prose. Analogously, one can say that since the dawn of history, mathematicians have been doing something of which they were not aware – semiotics. To combat the syndrome, the two authors of this book, together with other leading semioticians have founded a scholarly network at the Fields Institute for Research in Mathematical Sciences at the University of Toronto in 2012 to study the interface between semiotic theory and mathematics. This book is a product of an emerging semiotic mindset within mathematics itself.

In our first chapter we simply present an array of facts and historical anecdotes that show how semiotics, Modeling Systems Theory, and mathematical theory and practice are intrinsically intertwined, discussing some previous work in the semiotics of mathematics and delving into the main insights that can be gained by amalgamating semiotics with mathematical theorizing. The second chapter looks at the oppositional nature of basic mathematical ideas and models, discussing the general applications of structuralist opposition theory to mathematics. The third chapter deals with the pervasiveness and cognitive power of diagrammatic modeling in mathematics, utilizing Peirce's main notion of Existential Graphing as a backdrop to the discussion. The fourth chapter looks at the role of metaphor in mathematics, with a discussion of blending theory and metaform theory as frameworks for understanding the psychological processes involved in constructing many types of mathematical models. Finally, in the fifth chapter we will look at the relation between mathematical modeling and reality, that is, we will discuss whether or not mathematics is discovered or invented (or both). We conclude with a few general philosophical comments.

The reader might view many of the notions that we discuss in this book as "reformulations" of standard notions in mathematics and semiotics, pasting them together in our own particular way. We would like to suggest that this is so only in a coincidental way. Our claim is that by approaching mathematics from the standpoint of semiotics, it will put us in a better intellectual position to grasp and understand its basic structure, no matter how subjective the approach. We must also warn readers about what not to expect from this book. They will not find in it an in-depth treatment of mathematical or semiotic theory. Our goal is simply to show how the basic nature of mathematics can best be envisioned as a semiotic modeling activity. Finally, we should inform the reader that, since founding the Fields network, we have become enormously enthusiastic about the prospect of uniting semiotics and mathematics. We cannot but agree with Mark Twain, when he wrote that: "Intellectual work is misnamed; it is a pleasure, a dissipation, and is its own highest reward" (Twain 1889: 167).

Marcel Danesi and Mariana Bockarova, 2012

1. Mathematics, semiotics, and modeling systems theory

1.1. Introduction

In his groundbreaking 1962 study on the cognitive source of scientific theories, the American philosopher Max Black argued convincingly that the genesis of theoretical notions and frameworks in the sciences and mathematics was not solely the result of scientists deducing them from empirical observations or experimental results, but also, and primarily, the result of scientists making inferences and connections between facts, other theories, and even everyday experience (Black 1962). Indirectly, Black laid the foundations for a semiotic study of the nature of mathematics and science with his idea, radical for the era in which it was conceived.

Although Peirce referred to mathematics throughout his writings on sign theory, to the best of our knowledge, the earliest explicit suggestion that mathematics be studied from the standpoint of semiotics was made by Roman Jakobson. However, his entreaty has never really been fully realized or taken up systematically, either by semioticians or mathematicians, despite the fact that much has been written about the relation between the two fields, and between semiotics and mathematics education, since at least the mid-1970s (for example, Marcus 1975, 1980, 2003, 2010; Thom 1975, 2010; Rotman 1988; Varelas 1989; Reed 1994; MacNamara 1996; Radford, Grenier 1996; Radford et al. 2008; English 1997; Otte 1997; Anderson et al. 2000, 2003; Godino et al. 2007; Bockarova et al. 2012). There now exists intriguing evidence from the domains of education and psychology that semiotic concepts actually help explain fundamental aspects of how mathematics is learned (for example, Cho, Proctor 2007; Van der Schoot et al. 2009). The purpose of this opening chapter, therefore, is to take a look at the semiotics-mathematics interface in a general way, laying the groundwork for formulating specific questions and conceptualizations about the nature of mathematics from the viewpoint of semiotic theory.

In our view, the specific semiotic framework that would seem to apply best to mathematics is Modeling Systems Theory (MST), understood as a theory of how models are produced and used to carry out referential, cognitive, and practical tasks. The word *model* is an alteration of Latin *modulus* (an architectural unit of length), which became the basis of the word *modelle* in French to describe a set of plans for a building. Models are, in effect, "plans" devised by human beings to represent, symbolize, typify, illustrate, designate, simulate, copy, or substitute some real-world or imaginary referent. Model-making typifies all aspects of human knowledge. Architects make scale models of edifices, in order to visualize the structural and aesthetic components of building design; scientists utilize computer models of atomic and sub-atomic phenomena to explore the structure of invisible matter; mathematicians devise equations, proofs, and formulas to model temporal, spatial, and quantitative aspects of observed reality; and so on and so forth. As Jakob von Uexküll (1909) claimed, modeling is not unique to humans; it is found in all species. Current biosemiotic research is constantly confirming Uexküll's assertion, showing that modeling is a species-specific endowment (Sebeok, Danesi 2000; Kull 2001; Emmeche, Kull 2011), manifesting itself typically as a recurring behavior activity that serves some biological function in animals. Worker honey bees returning to the hive from foraging trips have the extraordinary capacity to inform the other bees in the hive about the location of a nectar or pollen cache with amazing accuracy through specific kinds of movement sequences. These sequences model the direction and distance of the location. When the food source is nearby, the bee moves in circles alternately to the left and to the right. When the food source is further away, the bee moves in a straight line while wagging its abdomen from side to side and then returning to its starting point. This kind of modeling system is instinctual, of course. As such, it is passed on genetically in the species. A lot of human modeling is also instinctual. But what sets it apart from modeling in all other species is its ability to transcend its biological paradigm and portray real world and imaginary events in creative and sophisticated ways, separate from any innate imprints.

1.2. Mathematics as a semiotic activity

The transition of mathematics from a practical counting, measuring, and generic problem-solving craft to a theoretical discipline is traced generally to the emergence of the method of proof beginning in the 500s BCE with Thales (c. 624–546 BCE), considered the founder of Greek philosophy and one of the seven wise men of Greece, and Pythagoras (c. 582–500 BCE). Pythagoras – or, more correctly, the Pythagoreans – were particularly instrumental in this transitional process after they proved using a logical system of demonstration that right triangles always had the same abstract structure, no matter what the length of their sides or the orientation of the triangles (Maor 2007). This method of proof was systematized and ensconced into mathematics by Euclid (c. 330–270 BCE) in his pivotal treatise, the *Elements*, around 300 BCE. Euclid's book was *ipso facto* the founding textbook of mathematics as a distinct theoretical discipline. Many of the definitions, demonstrations, and concepts in the *Elements* were taken from the Pythagoreans and other early mathematicians. The Pythagoreans divided the study of mathematics into four fields – arithmetic, geometry, astronomy, and music (called the *quadrivium* in the medieval period). In combination with the liberal arts of grammar, rhetoric, and logic (known as the *trivium*), the basis for establishing a curriculum for advanced learning was laid in the early academies of the western world and the first medieval universities.

The Pythagorean method turned practical knowledge into powerful theoretical knowledge. Around 2000 BCE, the ancient Egyptians, for instance, knew that knotting and stretching a rope into sides of 3, 4, and 5 units in length produced a right triangle, with 5 the longest side (the hypotenuse):

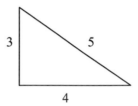

This was used by engineers, architects, and others to build structures, even though they had no explicit theoretical knowledge of why they were able to do so. The ancient Egyptians, for example, wanted to construct fields with square (90°) corners. Workmen knotted a loop of rope into 12 equal spaces and then they stretched the rope around three stakes to form a triangle, placing the stakes so that the triangle had sides of 3, 4, and 5 units. The Pythagoreans showed that this stretching pattern was not an isolated case, but rather a token of a more general pattern — the area of the square drawn on the hypotenuse equals the sum of the areas of the squares drawn on the other two sides. This relationship remains constant no matter what size we make the triangle. Thus, to produce a right triangle of any dimension, all one has to do is choose three stretches of rope that fit the pattern, which can be symbolized with the formula $c^2 = a^2 + b^2$, with c standing for the hypotenuse and a, b the other two sides. In other words, the Pythagoreans produced a *theorem* that encapsulates the features of this intrinsic pattern of right triangles. As the historian of science Jacob Bronowski (1973: 168) has insightfully written, we hardly realize today how important Pythagoras' theorem was to the birth of mathematics and the emergence of the scientific intellect:

> The theorem of Pythagoras remains the most important single theorem in the whole of mathematics. That seems a bold and extraordinary thing to say, yet it is not extravagant; because what Pythagoras established is a fundamental characterization of the space in which we move, and it is the first time that it is translated into numbers. And the exact fit of the numbers describes the exact laws that bind the universe. If space had a different symmetry the theorem would not be true.

As Bronowski notes, the theorem is not just a recipe of how to construct any right triangle; it is a model of space, since it tells us that certain spatial relations are the way they are because of a hidden structure inherent in them. The theorem makes this structure explicit, allowing us to study it on its own terms.

Actually, proofs of the theorem have been found in many parts of the ancient world (from China to Africa and the Middle East) long before the Pythagoreans (Strohmeier, Westbrook 1999; Bellos 2010: 53). The archeological discovery of a Babylonian method for finding the diagonal of a square suggests that the theorem was actually known 1,000 years before Pythagoras (Musser et al. 2006: 763). But Pythagoras was probably the first to prove that it holds for all right-angled triangles in a demonstrable way, even though, ironically, he left no written version of it (his proof is described through secondary sources). The common strategy involved in all known proofs of the theorem is the use of diagrams, which constitutes a fundamental modeling strategy in mathematics, as will be discussed in Chapter 3. Nobody knows for sure how Pythagoras proved his theorem. Many historians of mathematics believe (through the reports of ancient secondary sources) that it was a *dissection* proof, similar to the following one presented here for the sake of illustration. The first part of the proof is to draw a right triangle: $a^2 + b^2 = c^2$.

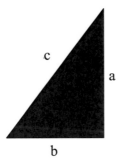

Then, we construct a square with length a + b (the sum of the lengths of the two sides of the triangle above). This is equivalent to joining four copies of the triangle together in the way shown below:

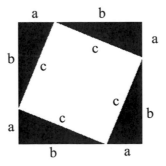

As can be seen, this configuration produces an internal square with area c^2. The area of the large square is $(a + b)^2$, which is equal to $a^2 + 2ab + b^2$. The area of any one triangle in the square is $1/2$ ab. There are four of them; so the overall area covered by the four triangles in the large square is: $4(1/2$ ab$) = 2ab$. If we subtract this from the area of the large square – $(a^2 + 2ab + b^2)$ – 2ab – we get $a^2 + b^2$. This corresponds to the area in white above, which is equal to c^2, the area of the internal square. The proof is now complete (see Posamentier, Hauptman 2010).

There are 370 known proofs of the theorem (Bellos 2010: 54). All of them make use of diagrams. It does not matter what particular form the diagram takes, so long as it shows the structural relation in a visual way. This shows that model-making is not fixed or predetermined, as for example bee dancing is. It is adaptive and subject to individual ingenuity. Among those who came up with original proofs are Leonardo da Vinci and American president James Garfield. Garfield published his proof in an 1876 issue of the *New England Journal of Education* while still a member of the House of Representatives.

A perusal of Euclid's *Elements* shows that symbolism and representational strategies played a major role in how he organized mathematical concepts and established theorems. Needless to say, Euclid made no direct reference to the emerging trend in Greek philosophy to adopt the concept of *semeion*, as put forward by Hippocrates and later extended to describe non-symptomatic signs. Among the first to differentiate between physical and conventional signs was Plato who was intrigued by the fact that a single word had the capacity to refer not to specific objects, but to objects that resemble each other in some identifiable way. As Plato realized, words reveal something remarkable about human understanding – namely, a propensity to unravel the essence of things, not just name and classify them as individual objects. The latter would, in any case, be impracticable because there would be as many words as there are things. Theorems in mathematics have the same power as words to unravel the essence of structure. Plato's illustrious pupil Aristotle took issue with his teacher's philosophy, arguing that words start out as practical strategies for naming singular things, not properties. It is only after we discover that certain things have similar properties that we start classifying them into categories. At such points of discovery, Aristotle argued, we create abstract words that allow us to bring together things that have the similar properties. In contrast to Plato's "mentalist" theory, Aristotle's theory is called "empirical". Both theories make sense and neither one can be proved or disproved. To this day, the debate between mentalists and empiricists rages on, indicating that it will probably never be resolved.

Eventually, the question arose as to whether or not there was any connection between natural and conventional signs. Among the first to discuss a possible relationship were the Stoics, who argued that conventional signs revealed something intrinsic about the nature of cognitive states in the same way that natural signs revealed something intrinsic about natural states. It was St. Augustine, the early Christian church father and philosopher, who was among the first to argue for a fundamental difference between the two in his *De doctrina christiana*. For St. Augustine, natural signs *(signa naturalia)* were distinct from conventional ones *(signa data)* because, he argued, they were a product of nature, and thus lacked intentionality. These included not only bodily symptoms, but also the rustling of leaves, the colors of plants, the signals that animals emit, and so on. Conventional signs, on the other hand, are the product of human intentions and of the need to understand the world. These include not only words, but also gestures and the many symbols that humans invent to serve their psychological, social, and communicative needs. Interestingly, as an early founder of sign theory, St. Augustine implored "the good Christian to beware of mathematicians and all those who make empty prophecies" (cited in Crilly 2011: 9). The reason why Augustine made this assertion is likely because in his era mathematics was closely connected to astrology and its murky practices.

Semiotics and mathematics parted ways early on and were never really united, as a look at the histories of both disciplines suggests. This does not mean that mathematicians never considered what they were doing as based on signs and symbols. Rather, it implies that they were unaware of the discipline of semiotics as a theoretical tool for analyzing how they conducted their theoretical activities. As mentioned, Peirce often alluded to mathematics in his studies and Jakobson emphasized the need for semioticians to apply their theoretical tools to mathematics. But the task of uniting the two disciplines has never really been undertaken, apart from some notable exceptions (Rotman 1988; Thom 2010; Marcus 2010). On the other hand, interest in semiotics on the part of mathematics educators has been burgeoning since the late 1980s, as witnessed by the number of studies, special issues of journals, monograph series, websites, and a host of books dealing with the semiotics-mathematics interface, either directly or indirectly (for example, Lave 1988; Varelas 1989; Reed 1994; Pimm 1995; Radford, Grenier 1996; MacNamara 1996; English 1997; Anderson et al. 2000, 2003).

1.3. Problem-solving

Anything can be used as a sign for representing number and space concepts, including body parts. Prehistoric people used their fingers, long before the advent of vocal language and the invention of words, to represent number concepts. Finger signs are *firstness* signs as Peirce called them (CP 1.23). People also used pebbles, knots in a cord, marks on wood or stone, and the like for the same representational purpose. From the latter the first number systems emerged. For example, around 3000 BCE the Egyptians developed a decimal system based on counting groups of ten (without place value); around 2100 BCE, the Babylonians invented a sexagesimal system based on counting groups of 60 – a system we still use to this day to mark the passage of time. These systems were used primarily to carry out practical tasks – surveying, constructing buildings, taxation, and so on. Geometry was also a practical measurement craft. Pythagoras was among the first who started looking for patterns within the systems and crafts themselves, apart from their practical uses. For example, he divided numbers into *odd* and *even* and *prime* and *composite*. Together with his theorem and his classifications, Pythagoras opened the door to the advent of mathematics as a theoretical discipline. Shortly after, Euclid founded the first school of mathematics in Alexandria, using the method of proof established by Thales and Pythagoras as the basis for establishing mathematical truths. Euclid finished his proofs with a concluding formula that was later translated into Latin as *Quod erat demonstrandum* ("which was to be demonstrated"). QED has remained the hallmark of mathematical method.

The term *mathematics* (from Greek *máthema* "knowledge, learning") was coined at about the same time that the term *semiotics* (from Greek *sêmeiotikos* "observant of signs") was being introduced into medicine by Hippocrates to designate the symptoms produced by the human body. Both disciplines were concerned with devising solutions to problems of specific kinds – mathematical and medical. It was Plato who coined the word *problem* to describe the activities that mathematicians carried out. Problem-solving was thus connected implicitly to scientific discovery and real-world activities. The English philosopher and scientist Roger Bacon (c. 1214–1292) was among the first to characterize mathematics explicitly in this way (cited in Kline 1969: 1):

> Neglect of mathematics works injury to all knowledge, since he who is ignorant of it cannot know the other sciences or the things of this world. And what is worse, men who are thus ignorant are unable to perceive their own ignorance and so do not seek a remedy.

From the outset, it was evident that successful problem-solving involved the ability to represent given facts in order to envision a solution strategy. Consider a simple problem in algebra:

> A cat weighs 20 pounds plus half of its own weight. How much does the cat weigh?

This problem can be solved easily by using a letter symbol, such as x, to represent the cat's unknown weight. Any other symbol (a dot, a line, and so on) could, of course, have been chosen to represent the weight. The problem tells us that the cat weighed a total of 20 pounds plus half of its own weight. This is represented by $1/2x$, of course. Now, the problem states that the cat's total weight (x) was 20 pounds plus $1/2x$ (half its own weight). This can be represented by the following equation: $x = 20 + 1/2x$. The symbols in the equation stand for the separate parts of the given statement. They are sign forms. The equation is a text, in the semiotic sense, translating the verbal text of the problem into a mathematical model of the problem. Solving the equation, we get x = 40. The equation is a perfect example of a *schema* model that mirrors iconically the structure of a balance scale, with x on the left-hand side of the scale and 20 + 1/2x on the right-hand side. The values that the sign forms stand for must be equal, otherwise the scale would not be balanced.

The early mathematicians solved such problems, not by using letter symbols and equations, but geometric diagrams. Diophantus (who lived in the third century CE) was perhaps the first to move away from this tradition, using symbols to stand for the facts in a problem, rather than diagrammatic forms. In some ways, the history of mathematics unfolds as a quest to find appropriate symbols to represent problems and to represent mathematical entities and forms. The signs + (plus) and – (minus) were introduced in 1489 by Johann Widman (1462–1498) and the equality sign (=) by Robert Recorde (c. 1510–1558) in 1557. William Oughtred (1575–1660) first used the symbol × for multiplication in 1631 and René Descartes (1596–1650) introduced superscript notation (a^n) at about the same time. John Wallis (1616–1703) introduced the symbol ∞ for infinity in 1655 and Thomas Harriot (1560–1621) the symbols > (greater than) and < (less than) in 1631. The Swiss mathematician Leonhard Euler (1707–1783) was responsible for the symbols Δ and f used in the theory of functions. He also introduced the use of the letter *i* for the square root of minus 1 in 1748. Without these sign forms mathematics would quickly become an impractical and unwieldy enterprise.

1.4. Number systems, numerical patterns, and classification

The most basic signs in mathematics are numerals, which stand for quantitative concepts (numbers). The *integers*, for instance, are signs that stand for holistic number concepts (rather than ones designating partiality), and are thus similar to the concepts encoded by mass nouns *(rice, information,* and so on). The sum or product of whole numbers always produces another whole number: 2 + 3 = 5. On the other hand, dividing whole numbers does not always produce another whole number, because division is akin to the process of partitioning something. As such, it is analogous to partitive structures in language (such as *some*). So, 2 divided by 3 will not produce a whole number. Rather, it produces a partitive number known of course as a *fraction*: 2/3. Various types of numeral sign systems have been used throughout history to represent such concepts. The connection between the numeral and its referent, once established, is bidirectional or binary – that is, one implies the other. The decimal system has prevailed for common use throughout most of the world because it is an efficient system for everyday number concepts. The binary system, on the other hand, is better adapted to computer systems, since computers store data using a simple on-off switch with 1 representing on and 0 off.

Counting with fingers or toes is an instinctive skill; but the use of specific types of numerals is culture-specific or a product of historical forces. Learning to connect numeral signs with counting patterns is an acquired skill subject to developmental processes, as the Swiss psychologist Jean Piaget showed this with a set of ingenious experiments with children (Piaget 1952). In one experiment, he showed a five-year-old child two matching sets consisting of six eggs placed in six egg-cups. Piaget then asked the child to tell him whether the eggs and egg-cups were equal or not. Invariably, the child would reply affirmatively. Piaget then took the eggs out of the cups, bunching them together, with the egg-cups left in the places where they were put previously, asking the child whether or not all the eggs could be put into the cups, one in each cup and none left over. The answer he typically received was that they could not. When he asked the child to count both eggs and cups, the child would correctly say that the number was equal. But when asked if there were as many eggs as cups, the child would again answer no. Piaget concluded that the child had not grasped the concept of numeration, which is not affected by changes in the positions of objects. To the semiotician, Piaget showed, in effect, that five-year-old children have not yet established in their minds the connection between numerals (signifiers) and their number referents (signifieds). As the psychologist Skemp (1971: 154) has pointed out, Piaget's work has shown that counting with numerals is a learned skill that overlaps with verbal development: "Counting is so much a part of the world around them that children learn to recite number-names not long after they learn to talk".

Numerals constitute a code. The code used commonly today is, as discussed, the decimal one (from Latin *decem* "ten"). It is based on ten signs called *digits* (meaning "fingers" in Latin). These can have verbal (*one, two, three...*) or visual form (*1, 2, 3, ...*). The distinguishing structural feature of this code is that the value (actual magnitude) to which any digit refers depends on the position it occupies in the numeral layout. The code is a symbolic one and thus is grasped later in development when symbol-using capacities emerge. Young children experience difficulty in reading decimal notation because its referential system cannot be indulged without grasping the underlying sign-formation rule – each digit in the numeral is 10 times greater than the value of the digit just to its right, and the total value that the numeral represents is determined by adding up the individual values of the digits that make it up. More technically, the value of a digit is determined by multiplying it by a successive power of ten (starting from the right side). Not unexpectedly, psychological research has shown that the ability to read numerals emerges at the same time as the ability to read words.

For a place-value numeral code to work, a sign showing that a certain place can be left empty (without value) is required. In the decimal system that sign is zero. The concept of zero was implicit in Babylonian mathematics, which simply left a blank space for it. The Chinese also left an empty space on their counting boards. There is archeological evidence that the Mayans had an actual symbol for zero by about 250 CE. But it was Indian scholars who invented the symbol we use today. It was used by the Indian mathematician Brahmagupta for the first time around 600 CE, not only as a place-holder, but more significantly, as a number itself. The word *zero* derives from *ziphirum*, a Latinized form of the Arabic word *sifr* which, in turn, is a translation of the Sanskrit word *śūnya* (void or empty). The zero was introduced to Europe along with the decimal system by Leonardo Fibonacci in his 1202 book called the *Liber Abaci*. Significantly, the perceptive Fibonacci starts off his book identifying zero as a sign (cited in Posamentier, Lehmann 2007: 11):

> The nine Indian figures are: 9 8 7 6 5 4 3 2 1. With these nine figures, and with the sign 0, which the Arabs call zephyr, any number whatsoever is written.

As Flood and Wilson (2011: 15) point out, the introduction into mathematics of a symbol for zero was a momentous event, because before that the context would be used to interpret a gap as a place. And that created confusions. More importantly, this simple device, as Brahmagupta realized, raises new questions about number concepts. In other words, the symbol led to further discovery in the code itself. This is a basic pattern in mathematics, as will be discussed throughout this book. It also shows the intrinsic interconnection between semiosis and mathematical knowledge. The invention of the zero symbol led to major discoveries, as zero became a digit in the system, not just a place-holder.

The zero concept is also an example of the oppositional value of numerical signs. Of course, in a binary system, there are only two signs, 0 and 1, and these by virtue of this fact will show opposition at every level (from their distinctive forms to their uses in numeration). But even in the decimal system the zero enters into oppositional relations of various kinds (as will be discussed in due course). As such, numerical signs are no different from any other kind of sign. As is widely known, basic to Saussure's (1916) plan for the study of language (and other sign systems) was the notion of *différence* ("difference, opposition") – namely, the view that sign structures do not take on meaning and function in isolation, but rather in differential relation to each other. A phonemic opposition, for instance, will show, among other things, that the initial consonants /k/ and /p/ of word pairs such as *kin* and *pin* are cues in English for establishing the differential meanings of the words. From such oppositions we can see, one or two features at a time, what makes the words unique in English, allowing us to pinpoint what they mean by virtue of how they are different from other words. He called the meanings the *valeur* that results from the oppositions. At this microlevel of linguistic structure, opposition is clearly a technique that allows us to sift meaningful signals out from the phonic stream that constitutes the chain of speech. We will revisit the theory of opposition in the next chapter, since it is our claim that mathematical oppositions are crucial to the construction of basic models. Suffice it to say here that an opposition is a default activity in mathematical theorizing, An opposition such as that between *even* and *odd* numbers, for example, is a basic one in mathematics, as is the *prime*-versus-*composite* one.

Opposition is an intrinsic feature of sign systems. But so too is combination and layout. So, a phoneme gains oppositional value because it is an element in a word or some other combinatory form. In mathematics, too, a digit has value only when it occurs in numerals, in other forms, or as an element in problems and layouts. When this happens, new insights emerge. As an example, consider the following layouts involving the number nine and the operations of multiplication and addition:

$2 \times 9 = 18$	and	$1 + 8 = 9$
$3 \times 9 = 27$	and	$2 + 7 = 9$
$4 \times 9 = 36$	and	$3 + 9 = 9$
$5 \times 9 = 45$	and	$4 + 9 = 9$
...		
$12 \times 9 = 108$	and	$1 + 0 + 8 = 9$
$123 \times 9 = 1107$	and	$1 + 1 + 0 + 7 = 9$
$1{,}245 \times 9 = 11{,}205$	and	$1 + 1 + 2 + 0 + 5 = 9$
$12{,}459 \times 9 = 112{,}131$	and	$1 + 1 + 2 + 1 + 3 + 1 = 9$
...		

These show that the digits of any multiple of 9, when added together, produce 9 (or a multiple of 9). Whatever its implications for mathematics, this pattern would likely never have been noticed without the above layout. In a fundamental sense, mathematics is the science of detecting patterns in layouts or other forms. The number of discoveries that this has brought about is astounding. It is instructive to consider a classic example of this principle from the seventeenth century, when the French mathematician Blaise Pascal decided to examine the binomial expansion of $(a + b)^n$ where n = {0, 1, 2, 3, 4, 5, ...} by laying out its successive expansions as follows:

$$(a + b)^0 = 1$$
$$(a + b)^1 = a + b$$
$$(a + b)^2 = a^2 + 2ab + b^2$$
$$(a + b)^3 = a^3 + 3a^2b + 3ab^2 + b^3$$
$$(a + b)^4 = a^4 + 4a^3b + 6a^2b^2 + 4ab^3 + b^4$$
$$(a + b)^5 = a^5 + 5a^4b + 10a^3b^2 + 10a^2b^3 + 5ab^4 + b^5$$
$$(a + b)^6 = a^6 + 6a^5b + 15a^4b^2 + 20a^3b^3 + 15a^2b^4 + 6ab^5 + b^6$$
$$(a + b)^7 = a^7 + 7a^6b + 21a^5b^2 + 35a^4b^3 + 35a^3b^4 + 21a^2b^5 + 7ab^6 + b^7$$
$$(a + b)^8 = a^8 + 8a^7b + 28a^6b^2 + 56a^5b^3 + 70a^4b^4 + 56a^3b^5 + 28a^2b^6 + 8ab^7 + b^8$$

If we look at the numerical coefficients in this layout, the shape of a triangle in outline form starts to jut out. That is what probably led Pascal to make his subsequent discoveries related to the binomial expansion. In the top row, the numerical coefficient is 1; in the row below the numerical coefficients of a and b are both 1; in the row below that the numerical coefficients are 1, 2, and 1; and so on. These can be highlighted as follows:

$$(a + b)^0 = \boxed{1}$$
$$(a + b)^1 = \boxed{1}a + \boxed{1}b$$
$$(a + b)^2 = \boxed{1}a^2 + \boxed{2}ab + \boxed{1}b^2$$
$$(a + b)^3 = \boxed{1}a^3 + \boxed{3}a^2b + \boxed{3}ab^2 + \boxed{1}b^3$$
$$(a + b)^4 = \boxed{1}a^4 + \boxed{4}a^3b + \boxed{6}a^2b^2 + \boxed{4}ab^3 + \boxed{1}b^4$$
$$(a + b)^5 = \boxed{1}a^5 + \boxed{5}a^4b + \boxed{10}a^3b^2 + \boxed{10}a^2b^3 + \boxed{5}ab^4 + \boxed{1}b^5$$

etc.

Manipulating the suggested triangular form a bit and deleting the literal coefficients produces the following triangle made up of the coefficients:

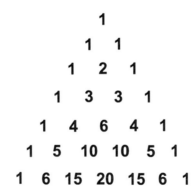

The diagram is, of course, only a slice of the resulting triangle, which has infinite dimensions. As it stands, it shows new and unexpected things. For example, each row begins and ends in 1 and the other numbers in a row are the sum of the two numbers above it: for example, each of the two 3's in the fourth row can be derived as the sum of the two numbers right above it (1, 2); the 6 in the third row can be derived from the sum of the two 3's above it; and so on.

That is just one of the hidden patterns in the triangle. Many more have been found. One of these is the fact that the diagonals unexpectedly produce the Fibonacci sequence (1, 1, 2, 3, 5, 8, ...), which will be discussed in the final chapter.

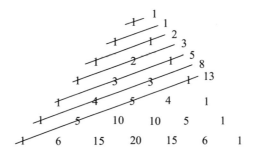

It is somewhat amazing to consider that a simple diagrammatic modeling of an algebraic expression produces so many new patterns on its own. Incidentally, when Isaac Newton was a student in his early twenties, he noted that the exponent of the binomial expression indicated the row of the triangle. For example, in the expansion of $(a + b)^4$, which is $a^4 + 4a^3b + 6a^2b^2 + 4ab^3 + b^4$, the numerical coefficients (1, 4, 6, 4, 1) coincide with the fourth row of numbers in the triangle; the coefficients in the expansion of $(a + b)^5$ coincide with the fifth row of numbers in the triangle; those in the expansion of $(a + b)^6$ with the sixth row; and so on.

The triangle is known, logically, as *Pascal's triangle*. For the sake of historical accuracy, however, it should be mentioned that historians found the same triangle in a Sanskrit text named the *Chandas Shastra*, written between 500 and 200 BCE (Stewart 2012: 113). The triangle also appears in a 1303 book, the *Precious Mirror of the Four Elements*, written by the Chinese mathematician Chu Shih-chieh. There is no evidence to suggest that Pascal was aware of any of these texts. These separate discoveries of the same mathematical pattern raise a host of interesting philosophical questions, which cannot be entertained here, although we will discuss some of them in the final chapter. The main one is whether mathematical truths and patterns exist "out there" and are discovered by different people at different times through their differential sign systems.

Now, many other discoveries and applications have resulted from the triangle – discoveries that would have been unimaginable without it. It is noticeable, for instance, in the calculation of probabilities. If a coin is tossed eight times in a row, one result might be all heads; another is all tails (H = head, T = tails):

(1) H H H H H H H H   ← *Eight heads thrown in a row*
(2) T T T T T T T T   ← *Eight tails thrown in a row*

Note that there are two outcomes in this case. Another result is seven heads and one tail. In this case, there are eight possible ways in which this can occur:

$$
\begin{aligned}
&(1)\ H\,H\,H\,H\,H\,H\,H\,T \\
&(2)\ H\,H\,H\,H\,H\,H\,T\,H \\
&(3)\ H\,H\,H\,H\,H\,T\,H\,H \\
&(4)\ H\,H\,H\,H\,T\,H\,H\,H \\
&(5)\ H\,H\,H\,T\,H\,H\,H\,H \\
&(6)\ H\,H\,T\,H\,H\,H\,H\,H \\
&(7)\ H\,T\,H\,H\,H\,H\,H\,H \\
&(8)\ T\,H\,H\,H\,H\,H\,H\,H
\end{aligned}
$$

For six heads and two tails, there are 28 outcomes (as readers may wish to verify for themselves); for five heads and three tails, there are 56 outcomes; and so on. Incredibly, these outcomes coincide with the numbers in the eighth row of Pascal's triangle (1, 8, 28, 56, 70, 56, 28, 8, 1). Referring to the outcomes in terms of tails, there is 1 outcome consisting of no tails; 8 outcomes of one tail; 28 outcomes of two tails; and so on. Altogether, the total number of possible outcomes of tails is:

$$1 + 8 + 28 + 56 + 70 + 56 + 28 + 8 + 1 = 256$$

In general, the outcomes of n tosses coincide with the numbers in the nth row of Pascal's triangle. Why this pattern is hidden in the triangle boggles the mind. Pascal is considered to be one of the founders of probability theory. It is unlikely that he would have come up with the theory without exploring the hidden patterns in his triangle. The triangle is an unexpected model of all kinds of new ideas. In a nutshell, this is why it can be claimed that mathematics is, *ipso facto*, a modeling system.

Number systems and the patterns they produce follow from the establishment of opposition classes (*even*-versus-*odd*, *prime*-versus-*composite*, and so on) form. These become *ipso facto* models of number concepts and their function as elements in distinct sets. For example, the *integer*-versus-*fraction* opposition has formed a basic class that has led to the discovery and use of the rational numbers. The relevant opposition here is *wholeness*-versus-*partiality*. An integer digit such as 1 represents a whole concept, such as a whole strip of paper, a digit such as 2 stands for two whole strips, and so on. To represent a part of the strip, such as one half or one-third, another type of numeral is needed, a fraction, in which two digits – one over the other – are used to indicate the type of partitioning involved: 1/2 (one half), 1/3 (one third), and so on. The fraction is a schema, an iconic sign-model of partition. The concept of fraction can also be modeled diagrammatically by cutting, say, a real strip of paper equally in half, in order to get two equal pieces:

Each of these is "one of two". If we cut the same strip into three equal parts, we will produce three equal pieces. Each of these is "one of three":

Oppositional classifications, as we have seen already, lead to further conceptualizations and derivative functions – a theme that will prevail throughout this book. In this case, the opposition introduces new numbers in themselves (fractions) that lead to expanded methods of computation. The diagrams actually show how the fraction sign-form relates to its value. As can be easily seen from the diagrams, 1/2 represents a larger piece than does 1/3. Thus, the fraction 1/2 is greater in value than 1/3. This is formalized as: 1/2 > 1/3. It is relevant to note that integers form the unmarked pole in the opposition, whereas fractions constitute the marked pole (see Chapter 2). As Jakobson (1942) and others found early on, learning marked oppositions comes later in development than does learning the unmarked ones. Psychological research shows, in fact, that without the diagrammatic models, many young children have a hard time believing that 1/3 is smaller than 1/2 because the denominator digit is larger in the former (Post et al. 1985).

Mathematics is *de natura* an oppositional modeling system. The concepts that basic oppositions generate allow us to envisage hidden structure, to compare structural patterns and their referential domains, and, as we have seen, to derive subsequent notions. These also lead to serendipitous discoveries. Oppositional classification is thus a guide to discovery, as we will also discuss in the next chapter. The *integer-versus-fraction* opposition allows us to understand a whole set of derivative notions and to expand the power of arithmetic to describe the world. The new number sign forms, *p/q*, with *p* the numerator and *q* the denominator, allow us now to define rational numbers and to carry out simple arithmetical operation. For example, if we compare the values of fractions with different numerators but identical denominators, such as 1/8 and 3/8, we can easily show diagrammatically that the numerator in this case indicates the larger value. We can use the diagram of a strip again cut into eight equal pieces. The numerator of 1/8 tells us that the fraction stands for "one" of those pieces, whereas the numerator of 3/8 tells us that it stands for three of the pieces.

Obviously, three is greater than two, so the value of the fraction 3/8 is greater than that of 1/8, as is, for example, also 2/8:

| 1/8 | 1/8 | 1/8 | 1/8 | 1/8 | 1/8 | 1/8 | 1/8 |

| 3/8 | | | 3/8 | | | 2/8 | |

When we compare fractions with different numerators and different denominators the reasoning becomes a little trickier, but still envisionable. Consider 2/7 and 3/4. The denominator of 2/7 tells us that a strip has been cut into seven equal pieces; the denominator indicates that there are two of these under consideration. The fraction stands for "two of the seven" pieces; and using similar reasoning the fraction 3/4 stands for "three of four" equal pieces in the same strip. Can these be compared? Let us draw appropriate diagrams. In the top figure, the strip has been cut into seven equal pieces. In the bottom one, the same strip has been cut into four equal pieces:

| 1/7 | 1/7 | 1/7 | 1/7 | 1/7 | 1/7 | 1/7 |

| 1/4 | 1/4 | 1/4 | 1/4 |

It can now be easily seen that one piece in the bottom strip is larger than one piece in the top one, hence 3/4 > 2/7. The point to be made here is that number systems, numerical patterns, and oppositional classifications are intrinsically interrelated and interdependent. Oppositions also lead to discoveries and further elaborations of arithmetical notions and concepts. A fraction can now be used as a model of the concept of *ratio*, represented, as indicated, with the general formula p/q, as long as q ≠ 0, because division by 0 is not allowed in mathematics. Actually, this new formula leads us to see that an integer can also be represented with it. In this particular case q has a constant value, namely, 1. So, for example, the first positive integer can be represented as 1/1, the second one as 2/1, and so on. The set of integers can now be shown to have the same structure of the fractions:

$$p/q = \{1/1, 2/1, 3/1, 4/1, ...\} = \{1, 2, 3, 4, ...\}$$

Because of this structural feature, revealed by the initial opposition, both the integers and the fractions can be classified together and called *rational numbers* (literally, numbers that can be represented as ratios). This new category suggests a further opposition: *rational-versus-irrational*. The latter can now be seen to constitute any number that cannot be represented with a p/q form. One such number is $\sqrt{2}$. Taking the square root of $\sqrt{16}$ produces an integer, namely 4. However, if we try to take the square root of $\sqrt{2}$, we will get a decimal with numbers after the point repeating at random:

$$\sqrt{2} = 1.4142...$$

It is historically interesting to note that the discovery of irrational numbers was produced by the Pythagorean theorem. The Pythagoreans considered this happenstance revelation a threat to their theory of numbers, which they associated with their theory of order – namely, that everything in the universe is based on rational numbers. The $\sqrt{2}$ was unexpected and unwanted. Yet it stared the Pythagoreans straight in the face each time they drew an isosceles right-angled triangle with equal sides of unit length. The length of its hypotenuse was the square root of the sum of $1^2 + 1^2$, or $\sqrt{2}$, a number that cannot be represented as the ratio of two integers, or as a finite or repeating decimal:

For the Pythagoreans, rational numbers had a "rightness" about them; irrational ones such as $\sqrt{2}$ did not. And yet, there they were, challenging the system of order that the Pythagoreans so strongly believed in. So disturbed were they that, as some stories recount, they "suppressed their knowledge of the irrationality of $\sqrt{2}$, and went to the length of killing one of their own colleagues for having committed the sin of letting the nasty information reach an outsider" (Ogilvy 1956: 15). The colleague is suspected to have been Hipassus of Metapontum (Bunt et al. 1976: 86; Aczel 2000: 19). As Crilly (2011: 40) puts it, the shock to the Pythagoreans came because if one attempts to measure the side of the triangle denoted by √2, one will never get the unit to finish flush along the side. Yet, there it is – it has length. But what length is it? Moreover, √2 is not the only irrational number. The set is infinite. To quote Crilly (2011: 41):

> After the square root of 2 failed to resolve into a fraction, a whole sequence of other square-root numbers presented a similar intractability – √3, √5, √6, √7, √8, √10, and so forth – missing out only the square roots of perfect squares like, 4, 9, 16, 25, etc.

The point to be made here is that a simple oppositional model of a number conceptualization (*wholeness*-versus-*partiality*) unexpectedly led to all kinds of new ideas and discoveries (wanted or not). This is a fundamental theme in this book. Suffice it to say at this point that discoveries are serendipitous events that result from the use of modeling systems (oppositional and otherwise) to understand the phenomena at hand. In many cases, the discoveries result from inferring what is already implicit in an existing model. It should also be noted, for the sake of historical accuracy, that the existence of irrationals and particularly of √2 was known to the Babylonians long before Pythagoras, suggesting that the basic wholeness-versus-*partiality* opposition may be a universal one. As Robin Flood and Raymond Wilson (2011: 19) note, the Babylonian notion was found on a tablet discovered by archeologists:

> A particularly unusual tablet, which illustrates the Mesopotamians' remarkable ability to calculate with great accuracy, depicts a square with its two diagonals and the sexagesimal numbers 30, 1;24,51,10 and 42;25,35. These numbers refer to the side of the square (of length 30), the square root of 2, and the diagonal (of length 30√2).

The discovery of irrationals was a milestone in the foundation and growth of mathematics as a modeling system. As mentioned several times, semiotically it showed that oppositional classificatory models generate unexpected new models, as irrationals quickly became themselves new models of mathematical structure opening up new vistas that would have otherwise been inconceivable.

An analogous story can be told about the discovery of imaginary numbers. These came about as a consequence of using square numbers and irrationals for the solution of quadratic equations, among other things. At some point the following quadratic equation caught someone's attention:

$$x^2 = -1$$

Taking the square root of both sides produces:

$$x = \sqrt{-1}$$

What kind of number is the square root of a negative number? Squaring a negative number produces a positive number, never a negative one:

$$-2^2 \ = \ -2 \times -2 = +4$$
$$-3^2 \ = \ -3 \times -3 = +9$$
$$-4^2 \ = \ -4 \times -4 = +16$$
$$\dots$$
$$-n^2 \ = \ +n^2$$

So, what is $\sqrt{-1}$? Not knowing what to call it, mathematicians referred to it, logically enough, as an *imaginary number*. Like $\sqrt{2}$ for the Pythagoreans, it did not fit any known schema at the time of mathematical structure and taxonomies. It simply surfaced in the domain of quadratic equations. As it has turned out, imaginary numbers have led themselves to an expansion of the number system, to the generation of new numerical patterns, and so on and so forth. An imaginary number opened up a new classificatory opposition, *real*-versus-*imaginary*, and was inserted into a new conceptualization of numbers known as *complex numbers*, which are defined as numbers having the form $a + bi$, where a and b are real numbers and i the square root of -1. The name *imaginary* is due to Descartes in the seventeenth century and the use of the symbol $i = \sqrt{-1}$ is due to Euler in the eighteenth century. Incredibly, complex numbers turn out to have many applications. They are used, for instance, to describe electric circuits and electromagnetic radiation and they are fundamental to quantum theory in physics. Stewart (2012: 318) puts it as follows:

> Who would have predicted in the fifteenth century that a baffling, apparently impossible number, stumbled upon while solving algebra problems, would be indelibly linked to the even more baffling and impossible world of quantum physics – let alone that this would pave the road to miraculous devices that can solve a million algebra problems every second, and let us instantly be seen and heard by friends on the other side of the planet?

Imaginary numbers are mentioned in previous writing, although they are not named as such. One is the work of the great twelfth-century Indian mathematician Bhaskara (1114–1185), where the idea of an imaginary number is discussed. A little later, in his treatise on equations, *Ars magna* (The Great Art), Italian mathematician Girolamo Cardano (1501–1576) introduced the concept of the square root of a negative number to a European audience. His compatriot Raffaello Bombelli (1526–1572) considered complex numbers to be solutions of cubic equations – equations in which one of the variables is cubed (x^3). Bombelli was the first to provide a consistent theory of complex numbers in his *Algebra* of 1572, specifying how the four operations (addition, subtraction, multiplication, and division) could be performed with complex numbers. It was Jean-Robert Argand (1768–1822) who gave complex numbers their first true comprehensive treatment, followed by Carl Friedrich Gauss's exhaustive study of such numbers. Interestingly, every real number can be represented as a complex number by simply letting the imaginary part be equal 0. So, for example, 5 can be represented as follows:

$$a + bi = 5 + 0i = 5$$

There are many equations that have no real number solutions, such as $x^2 + 1 = 0$. However, if x is allowed to be complex, the equation has the solutions $x = \pm i$. For example, $(\sqrt{-4})(\sqrt{-9})$ does not equal $\sqrt{36}$ or 6. Rather, it equals $(2i)(3i) = 6i^2 = 6(-1) = -6$. Such results led to the subsequent discovery of so-called quaternions $(a + bi + cj + dk)$ and octonions $(a + bi + cj + dk + ep + fq + gr + hs)$, with i, j, k, p, q, r, s representing imaginary numbers. Incredibly, such numbers have specific kinds of geometrical representations and correspond to different dimensions beyond three. As it has turned out quaternions have been used in "navigation, robotics and in the design of computer games" (Crilly 2011: 53).

The discovery and incorporation of imaginary numbers into mathematics have led to a remarkable equation that combines two enigmatic numbers, *e* and π, with i:

$$e^{i\pi} = \cos \pi + i \sin \pi = -1$$

Modifying this equation by plugging values in it produces yet another amazing formula:

$$e^{i\pi} + 1 = 0$$

As Banks (1999: 72) puts it, this is a remarkable result because "Here is an equation that uniquely relates the five most important numbers in all of mathematics: *e*, *i*, π, 1, and 0". And graphing equation numerous hidden patterns are revealed that would otherwise have gone unnoticed. As Stewart (2012: 84) observes, the very existence of this equation has not revealed all its hidden implications. For this reason, mathematicians refer to it as "beautiful":

> The imaginary number i unites the two most remarkable numbers in mathematics, e and π, in a single elegant equation. If you've never seen this before, and have any mathematical sensitivity, the hairs on your neck raise and prickles run down your spine. This equation, attributed to Euler, regularly comes top of the list in polls for the most beautiful equation in mathematics.

Episodes such as the ones discussed above typify the development of mathematics as a theoretical discipline. Connected to the use of imaginary numbers, and negative numbers generally, is the concept of the *absolute value* of any number n, which is shown as |n|. This is the value of the number regardless of its sign. Its use is not as spectacular as is the concept of imaginary number, but it is vital to the functioning of the number system as a representational one.

Number	Absolute Value		
0	$|0|$	=	0
15	$|15|$	=	15
21	$|21|$	=	21
1/2	$|1/2|$	=	1/2
−15	$|-15|$	=	15
−199	$|-199|$	=	199
−4/5	$|-4/5|$	=	4/5

The absolute value tells us what a number's actual distance from 0 is, without regard as to whether it is to the left or to the right of 0 on a number line. So, $|5|$ and $|-5|$ have the same absolute value (five), and this means simply that they are equidistant from 0. Obviously, 5 lies to the right of 0 and –5 to the left. Now, the interesting aspect of this simple idea is that the notion allows one to carry out arithmetical operations with signed numbers effortlessly. If the signs are the same, one can add the absolute values of the numbers and give the sum the common sign: (+3) + (+8) = (+11) and (−3) + (−8) = (−11). If the signs are different, one can subtract the smaller absolute value from the larger absolute value, and give the result the sign of the number with the larger absolute value: (+3) + (−8) = (−5) and (−3) + (+8) = (+5). In effect, the concept of absolute value has been found to facilitate many areas of mathematical practice; it corresponds, clearly, to the Saussurean *value*-versus-*substance*, linking mathematics to language in a meta-structural way – that is, in a way that shows that the oppositional and overall structural patterns in both are analogous or isomorphic, if not identical.

Absolute values, irrational numbers, imaginary numbers, and so on and so forth are the result of modeling. This is how fundamental mathematical ideas are interrelated and discovered. As mathematicians use existing models of structure, they come across new ideas and new possibilities. This leads to further discoveries and elaborations. Consider, as one more example of this general semiotic principle, the number represented by the alphabet character *e*. This is defined as the limit of the expression $(1 + 1/n)^n$ as n becomes large without bound. Its limiting value is approximately 2.7182818285. It is sometimes called "Euler's number" because Euler was the one who popularized it. It was Charles Hermite in the nineteenth century who proved that the number was special – not irrational but transcendental, that is, not the solution of any algebraic equation (Crilly 2011: 42). Now, one may ask, what possible connection does this number have with other areas of mathematics or with anything in the world? As it turns out, it forms the base of natural logarithms; it appears in equations describing growth and change; it surfaces in formulas for curves; it crops up frequently in probability theory; and it appears in formulas for calculating compound interest.

Another example of this principle is Descartes' analytic geometry. Descartes simply took two number lines and made them cross at right angles. From this he was able to plot points in the resulting plane. This then revealed how the fields of arithmetic, algebra, and geometry were interrelated, explaining why Descartes called his amalgamation *analytic geometry*. A number line is itself a rudimentary geometric model that shows the continuity between positive and negative numbers and a one-to-one correspondence between a specific number and a specific point on the line. By making two such lines intersect at right angles new vistas are opened up. The horizontal line is called the x-axis, the vertical one the y-axis, and their point of intersection the origin. This simple diagram allows the plane to be mapped as a system of points that are determined by their positions in relation to the two axes:

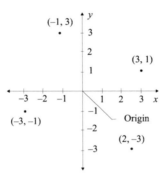

Each point in the plane can be identified by such ordered pairs of coordinates. The first one is the x-coordinate, expressing the distance to the left or right of the y-axis. If it is positive it is to the right; if it is negative it is to the left. The second one is the y-coordinate, expressing the distance above or below the x-axis. If it is positive, it is above; if it is negative, it is below. The point of intersection of the two axes, O, is called the origin. Its coordinates are, of course, (0, 0). The point (3, 1), for instance, is defined by its location with respect to the axes: that is, it is located three equally-calibrated units to the right of the y-axis and one unit up from the x-axis; the point (2, –3) is located two units to the right of the y-axis and three units down from the x-axis; the point (–1, 3) is located one unit to the left of the y-axis and three units up from the x-axis; and the point (–3, –1) is located three units to the left of the y-axis and one unit down from the x-axis. In this modeling system, points (p_n) are defined in relation to each other: p_n (x_n, y_n).

The number of subsequent discoveries show that this simple play with intersecting number lines is immeasurable, not to mention the use of analytic geometry in science and engineering. Suffice it to say that it reveals how discoveries are made. This whole line of discussion leads to a philosophical consideration of the Platonic versus constructivist views of mathematics. Do we discover mathematics or do we invent it and then discover that it works? The Platonic view was weakened by Gödel's (1931) incompleteness theorem. So, if mathematics is an invention, why does it lead to demonstrable discovery? René Thom (1975) referred to the fortuitous discoveries in mathematics as "catastrophes" in the sense of events that subvert or overturn existing knowledge (Wildgen, Brandt 2010). Thom also names the process of discovery as "semiogenesis" or the emergence of meaningful or "pregnant" forms (in his terminology) within symbol systems themselves. Forms emerge catastrophically, that is by happenstance through contemplation and manipulation of previous forms. As this goes on, every once in a while, a catastrophe occurs that leads to new insights as it disrupts the previous system.

Does catastrophic discovery exist as a cross-species ability? Actually, this question was approached by the Alexandrian geometer Pappus (c. 290–350 CE). Pappus contemplated a truly fascinating problem in geometry: What is the most efficient way to tile a floor? He concluded that there are three ways to do so with regular polygons of the same type, shown below (Flood, Wilson 2011: 36) – that is, with equal squares (or rectangles), equilateral triangles, or regular hexagons:

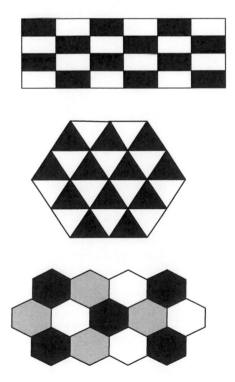

Now, Pappus went on to observe that bees instinctively understood that the last pattern was the best one for their honeycombs. In other words, bees chose the pattern with the most angles, the hexagonal pattern, because it holds more honey than the other ones. This is a truly mind-boggling consideration, raising a whole series of questions that cannot be taken into account here. As Uexküll (1909) might have put it, the internal modeling system of bees is well adapted to understanding the external world that they face, producing instinctual knowledge of that world. On this topic, the great Johannes Kepler (1571–1630) made the following observation (cited in Banks 1999: 19):

> What purpose had God in putting these canons of architecture into the bees? Three possibilities can be imagined. The hexagon is the roomiest of the three plane-filling figures (triangle, square, hexagon); the hexagon best suits the tender bodies of the bees; also labour is saved in making walls which are shared by two; labour would be wasted in making circular cells with gaps.

The modeling system of bees is designed to do exactly what they end up doing. Hexagonal structure is also found in nature. It has been found, for instance, that the molecular structure of snowflakes and ice crystals reveals a hexagonal form. The astonishing thought is that the hexagon is also produced by humans for a whole set of non-instinctual reasons. This overlap of cross-species modeling systems is an area that is starting to grab the attention of biologists, mathematicians, and semioticians alike. As Banks (1993: 19) also puts it: "It is noteworthy that currently mathematicians and scientists are devoting much more attention to research on topics advocated by Kepler".

1.5. Modeling Systems Theory

As the foregoing discussion implies, it is obvious understanding mathematics implies understanding how mathematicians use modeling strategies such as diagrams to carry out their proofs, their computations, their notations, and so on. This unites mathematics and semiotics epistemologically, even though this has never really been articulated explicitly in such a way, except for notable exceptions (Thom 1975, 2010; Marcus 2010). The goal of both disciplines has always been to figure out how sign forms are constituted and how they encode referents leading to knowledge of the world through their own internal action. Both aim to determine the nature, basis, and extent of knowledge, exploring the various ways of knowing, the nature of truth, and the relationships between knowledge, belief, and semiosis. Epistemological theories of correspondence, pragmatics, and coherence are implicit in the *modus operandi* of both sciences. The correspondence theory holds that an idea is true if it corresponds to the facts or reality. The pragmatic theory maintains that an idea is true if it works or settles the problem it deals with. The coherence theory states that truth is a matter of degree and that an idea is true to the extent to which it coheres (fits together) with other ideas that one holds. As we have seen in this chapter, mathematical models often correspond to the facts of reality, even when they are not constructed for this reason. And, of course, many models are born pragmatically to make computation easier or to facilitate some procedure. But, as we have seen, the models contain hidden information that leads to further insights which, in turn, often correspond to reality. Finally, coherence seems to be the rule in mathematical modeling, as we will discuss in the final chapter of this book. Overall, it can be claimed that mathematics is a particular kind of modeling system that allows for the exploration of reality on its own terms. Thus, before proceeding, it is worth going through the version of MST that we adopt here to make this claim.

Modeling is the use of sign forms to represent the world. A model is an interpretation of that world. It can be simulative (iconic), indexical (relational), or symbolic (based on conventions). Modeling typifies all aspects of human intellectual, aesthetic, and social life. Explorers will draft a map of the territory they anticipate investigating to guide them in their search. A scientist will produce a diagrammatic model of atoms and subatomic particles in order to get a mental look at them and then devise equations to describe their structure. Models turn raw information (the umwelt) into knowledge schemata (the *Innenwelt*). Sebeok and Danesi (2000) propose a typology of four basic model types in order to make MST applicable to various domains outside of semiotics proper: *singularized, composite, cohesive,* and *connective.* In traditional semiotic theory *singularized* models are single *signs.* A singularized model is a form (a digit sign, a symbol such as an arrow, and so on) that has been made to represent a singular (unitary) referent or referential domain. As we have seen, a digit form representing a whole concept is an integer (1, 2, 3, …). The digit form representing a partial concept (fraction) is constructed to show how a partition has occurred (1/2, 2/3, 3/4, …). In other words, the actual form of the digit suggests the nature of the referent. This is why a fraction has a p/q form, suggesting partiality in an iconic way. The fractional digit is therefore a model of partiality. *Composite* models are *texts* in the semiotic sense. They are models that have been made to represent various aspects of a referent or referential domain in a composite (combinatory) manner. An equation or a formula is a composite model. For example, the Pythagorean formula $c^2 = a^2 + b^2$ is a model of how certain sets of integers relate to each other. It is also a textual descriptor of right triangles. A *cohesive* model is a code or system, such as the system of integers, rational numbers, complex numbers, and so on. This kind of modeling strategy reveals the tendency to perceive certain forms in some *cohesive* fashion and thus classify them accordingly. A cohesive system can thus be defined as a system of forms that allows for the modeling of referents perceived to share common traits. Finally, a *connective* model is one that blends other models on-tologically to produce a new one that is perceived to have a common essence with previous ones. Metaphors and metoynyms, for example, are connective models, linking one type of referent to another to flesh out common essences in both. These four types of modeling strategies are not mutually exclusive. They are interdependent – singularized models go into the make-up of composite ones which, in turn, are dependent upon the forms that cohesive systems make available, and so on.

The function of singularized modeling, as mentioned, is to represent single, unitary referents. Numerals do exactly that. Roman numerals such as I, II, and III are iconic models because they stand for their referents in a directly visual way (one stroke = one unit, two strokes = two units, three strokes = three units). A number line is an indexical model, showing where a digit can be located. A singularized form that is constructed in terms of some historical practice or some conventional custom is a symbol. The use of a letter to represent an unknown in algebra is an example of a symbolic singularized form. Triangles too are singularized forms, referring to a specific type of configuration and con-stituting a unitary referent. The case of the triangle can be used to distinguish between *structure, form*, and *model*. A triangle is a triangle if it has a given structure – three sides joined to produce a plane figure with three angles. This structure can take on various forms (scalene, isosceles, equilateral, and so on). Each form is a token of the general structure, showing how the structure can be manipulated to produce variations:

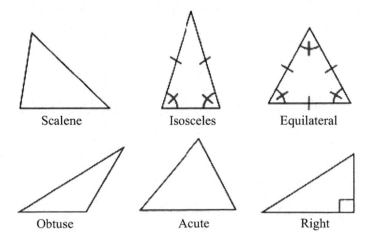

Scalene Isosceles Equilateral

Obtuse Acute Right

Any one of these forms can become a model of something. For example, the obtuse triangle can be used as a model of a sailboat or of one of its sails. As such it gives the builder a visual idea of what to build. The model can, of course, be made much more sophisticated with the addition of detail, but the outline will remain. The builder can then turn the visual model into a set of mathematical statements and forms (measurements, equations, and so on), which take the place of the visual model and perform the same modeling task in another way.

Composite modeling is the activity of representing complex (non-unitary) referents by combining various forms in some specifiable way. Diagrams and equations are examples of composite forms. These are constructed with singularized forms that fit together structurally, but which are, as a whole, different from any of their constituent forms taken individually. In analogy to atomic theory, a singularized model can be compared to an atom and a composite one to a molecule made up of individual atoms, but constituting a physical form in its own right, that is, composite models incorporate the structural properties of the individual forms with which they are constructed, but are not conceptually equivalent to the aggregate of their referents. Take the Pythagorean equation as a case in point: $c^2 = a^2 + b^2$. This equation is a model of a right triangle; it shows which integers can be used to constitute a triangular shape; and so on. It is made up of constituents in a structured way, where the hypotenuse (c) is put into a defined relation with the other sides.

As mentioned, a cohesive modeling system is known in traditional semiotic theory as a code, and in mathematics as a set. It can be defined as a system providing particular types of forms that can be used in various ways and for diverse modeling purposes. Generally speaking, for some particular representational need there is an optimum code or set of codes that can be deployed. Thus, the integers form a cohesive system that allows for the mapping of holistic referents. The laws of exponents form a cohesive system for working with exponential numbers and their applications.

Connective modeling is the result of a reasoning process that is now called *blending*. The ever-burgeoning literature on what has come to be known as *conceptual metaphor theory* (Lakoff, Johnson 1980, 1999; Lakoff 1987; Johnson 1987; Gibbs 1994; Goatley 1997; Fauconnier, Turner 2002; Gibbs, Colston 2012) has made it obvious that such modeling permeates cognition and communicative behavior. A connective form in language, for example, results when an abstract referent is represented in terms of a concrete one. The conceptual metaphor *thinking is seeing*, for instance, is a connective form because it delivers the abstract concept of "thinking" in terms of the physiological processes associated with "seeing". This underlies utterances such as: "I do not *see* what possible use your *ideas* might have"; "I can't quite *visualize* what that new *idea* is all about"; "Just *look at* her new *theory*; it is really something"; "I *view* that *idea* differently from you"; and so on.

A specific metaphorical statement uttered in a discourse situation is now construable as a particular externalization of a connective model, which Lakoff and Johnson have called an *idealized cognitive model*. So, when we hear people using such statements as those cited above, it is obvious that they are not manifestations of isolated, self-contained metaphorical creations, but, rather, specific uses of a model connecting thinking with seeing.

A connective form may also be the product of metonymic modeling. Metonymy entails the use of an entity to refer to another that is related to it. A metonymic model results when part of a domain starts being used to represent the whole domain (Lakoff, Johnson 1980: 35–40): "She likes to read Dostoyevski (= the writings of Dostoyevski)"; "He's in dance (= the dancing profession)"; "My mom frowns on blue jeans (= the wearing of blue jeans)"; "Only new wheels will satisfy him (= car)"; and so on. Each one of these constitutes an externalization of a metonymically-derived model: the author is his or her work, an activity of a profession is the profession, a clothing item represents a lifestyle, a part of an object represents the entire object, and so on.

In mathematics, many models emerge as the result of connective thinking. For example, the imaginary numbers have taken a place in mathematics after they were connected to the real numbers to produce complex numerals. In a recent lecture, Lakoff (2011) outlined a fascinating analysis of how mathematicians form their proofs through such connective thinking. He illustrated his argument with Kurt Gödel's famous indeterminacy proof, claiming that metaphorical thinking guided the way in which Gödel (1931) laid out his proof. The preexistent model that influenced Gödel was Georg Cantor's famous proof that there are more irrationals than rationals, which was guided by a simple metaphor – the familiar geometrical image of a diagonal. This metaphor manifests itself in Cantor's well-known diagrammatic layout – a square array of numbers through which a diagonalized subset comes into view to show the tenability of the theorem (as will be discussed subsequently in this book). Gödel's proof is the result of an analogous metaphorical argumentation, showing that within any formal logical system there are results that are undecidable through a diagonalized form of argument. Lakoff claimed that Gödel was clearly guided by Cantor's diagrammatic model, leading him to imagine three metaphors of his own. Lakoff calls the first one the *Gödel Number of a Symbol*, which can be seen in Gödel's argument that a symbol (in any formal system) is the corresponding number in the Cantorian one-to-one matching system (whereby any two sets of symbols can be put into a one-to-one relation). Gödel's second metaphor, which Lakoff calls the *Gödel Number of a Symbol in a Sequence*, was his argument that the nth symbol in a sequence is the nth prime raised to the power of the Gödel Number of the Symbol. And Gödel's third metaphor, which Lakoff calls *Gödel's Central Metaphor*, was Gödel's demonstration that a symbol sequence is the product of the Gödel numbers of the symbols in the sequence. For Lakoff such reasoning reveals that the brain identifies two distinct entities in different neural maps as the same entity in a third neural map. The three maps together constitute what he calls a *blend*. This will be discussed in detail in Chapter 4. Suffice it to say here that the whole of Lakoff's argument is ensconced in MST and the principle of discovery enunciated in this chapter – one form suggesting another suggesting another still, and so on – can also be enlisted to explain proof in logic and mathematics.

It is relevant to note that mathematicians have started to take serious note of the concept of modeling. The calculus, for instance, is now defined by mathematicians themselves as a modeling system of change and flux in the world, both encoding it and predicting it in its simple differential and integral equations. Crilly (2011: 84) describes the situation in mathematics as following:

> The field of differential equations is huge, and besides mathematicians it attracts physicists involved with physical theories, chemists interested in diseases and how fast they are spread. These are studied within the framework of *mathematical modeling* [emphasis ours], where simplifying assumptions are made in order to understand a process. Many areas where the Calculus is applied involve quantities with more than one variable, such as space *and* time.

Crilly (2011: 152) goes on to state that a "mathematical model is a way of describing a real-life situation in mathematical language, turning it into the vocabulary of variables and equations". When looked at globally, it can be seen that there are two kinds of mathematical models: (1) those intended to model reality and (2) those intended to model pattern within mathematics itself. In this book, type (1) models will be called *factual* and type 2 *artifactual*. Perhaps the most intriguing aspect of mathematical modeling is the fact that the two types of models often converge, producing scientific models of something real, as we have seen various times above.

The Pythagorean theorem can be used again to show the relation between artifactual and factual modeling. The original theorem was designed to generalize the fact that a rope can be stretched to form a triangle of sides 3, 4, and 5. The theorem was born as a factual model, showing how this stretching pattern holds generally. As we saw, it produced a derivative notion, namely that various triplets of whole numbers fit the Pythagorean formula of $a^2 + b^2 = c^2$, others do not. For example, {3, 4, 5}, {6, 8, 10}, and {5, 12, 13} are triples that fit the equation. As it turns out there are an infinite number of such triples. The equation $a^2 + b^2 = c^2$ is thus a derivative or artifactual model. As Stewart (2012: 7) suggests, these triples were known before Pythagoras, having been discovered in an ancient table of numbers (Plimpton 322):

It is a table of numbers, with four columns and 15 rows. The final column just lists the row number, from 1 to 15. In 1945 historians of science Otto Neugebauer and Abraham Sachs noticed that in each row, the square of the number (say c) in the third column, minus the square of the number (say b) in the second column, is itself a square (say a). It follows that $a^2 + b^2 = c^2$, so the table appears to record Pythagorean triples. At least, this is the case provided four apparent errors are corrected. However, it is not absolutely certain that Plimpton 322 has anything to do with Pythagorean triples, and even if it does, it might just have been a convenient list of triangles whose areas were easy to calculate. These could then be assembled to yield good approximations to other triangles and other shapes, perhaps for land measurement.

In other words, the triples were known practically, but their theoretical implications were not. This derived model of numbers has itself led to further work and discovery within mathematics. One of the most emblematic episodes of this occurred in 1637, when French mathematician Pierre de Fermat (1601–1665) wrote in the margin of his copy of Diophantus's *Arithmetica* that there is no whole-number solution of $a^n + b^n = c^n$ if n is greater than 2. In other words, aside from the Pythagorean triples, produced when n = 2, the equation does not always hold. Fermat wrote in the same margin that he had found a straightforward proof of this fact, but that there was not enough room to write it down. Fermat never published his purported proof and no proof of "Fermat's last theorem" was found for more than 350 years. In 1993, British mathematician Andrew Wiles announced that he had proved the theorem. Wiles published his complete proof, with certain corrections, in 1995 (Wiles 1995). The proof would not have been imaginable without the original Pythagorean theorem. As Stewart (2012: 14) aptly puts it, "Pythagoras's theorem, then, is important in its own right, but it exerts even more influence through its generalizations". As a modeling form, it has revealed the power of the human mind to infer things from previous ideas and then to apply them to discoveries and activities that would otherwise have never come to consciousness.

The calculus is another example of a derived or artifactual modeling system fusing with factual scientific models – it is used by engineers, physicists, and other scientists to develop models of their own and solve practical problems in their fields. For instance, the laws of aerodynamics are describable in terms of the calculus. An airplane designer can use these to calculate the changing forces that affect an airplane during flight. The differential calculus began with questions about the speed of moving objects: How fast does a stone fall two seconds after it has been dropped from a higher location? How fast is the earth moving around the sun on a specific date? The other branch of calculus, integral calculus, was invented to answer a very different kind of question: What is the area of a shape with curved sides? It is amazing to contemplate that the answers to these scientific question are inherent in the modeling system known as the calculus.

Ultimately, the calculus itself derives not from factual modeling of the world, but by considering artifactual paradoxes, such as those formulated by Zeno of Elea (c. 490–430 BCE). Zeno tried to prove that motion, change, and plurality (reality consisting of many substances) are impossible. He used *reductio ad absurdum* (reduction to the absurd) arguments, which imply deriving impossible conclusions from the opinions of his opponents. Zeno devised at least forty such arguments, but only eight have survived. His four paradoxes concerning motion make up his most famous surviving ones. In one of these, Zeno argued that a runner can never reach the end of a race course. He argued this by stating that the runner first completes half of the course, then half of the remaining distance, and so on infinitely without ever reaching the end. So, if the length of the race course is represented by a line, with unit length, the successive stages of the runner's location can be shown as follows:

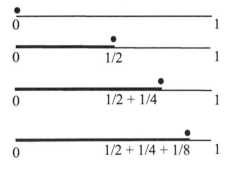

The successive stages of the race form an infinite series with each term in it half of the previous one:

1/2 + 1/4 + 1/8 + 1/16 + 1/32 + ...

Zeno argued that the runner would never cross the finish line, even though we know that he actually does (see Salmon 1970 and Mazur 2008 for detailed discussions). Zeno's apparently simple argument raised profound issues about time, space, and infinity. It led to the concept of *limits* which, in turn, inspired the invention of the calculus. Zeno's paradoxes treat distance and time as if they can be segmented into infinitely smaller parts or points. The answer of the calculus is that there are no such points in the flux of things.

The calculus also finds its source in the solution of problems of area, such as the one devised by Archimedes who used an ingenious diagram and idea to calculate the value of π. His diagram involves inscribing an infinite number of polygons in a circle (Askew, Ebbutt 2011: 73). Consider a circle with radius equal to 1/2. We inscribe a regular hexagon in it, bisecting the arcs into which the six vertices of the hexagon divide the circle:

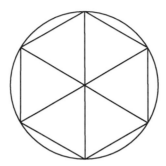

The perimeter can easily be determined (with basic geometry). Archimedes continued in this way obtaining successively regular polygons of 12, 24, 48, and 96 sides. The perimeter of each of these can also be determined, even if with some effort. As the number of sides in an inscribed polygon increases, the perimeter gets closer in value to the circumference of the circle. What is the circumference of the circle? It is π. Actually, the perimeter of a polygon with 96 sides falls short, but for the purpose of getting an estimation it is good enough, turning out to be between 3 and 10/71 and 3 and 1/7. Six hundred years after Archimedes, around 400 CE in a set of Indian manuscripts, called the *Siddhantas* ("Systems of Astronomy"), the value for π is given as 3.1416. The Chinese mathematicians also searched for a value for π. Lin Hui (c. 250 CE) used a variation of Archimedes' method, inscribing a polygon of 3,072 sides to determine π as equal to 3.14159.

Examples such as these bring out the power of models and the role of history to stimulate new ideas and to bring about new inventions and discoveries. In our view, MST (in any of its forms) can be used profitably to help us understand the connection between semiosis and reality and, in the case of mathematics, why mathematical models are so powerful as descriptors of reality. As Kull (2012: 334) has recently written, mathematics is a special case of semiotics, attempting to remove contradictions from its language. But in so doing it leads to further discoveries, because contradiction is the heart and soul of semiosis.

2. Opposition theory in mathematical modeling

2.1. Introduction

In the previous chapter we discussed the crucial role that oppositional thinking plays in the formation of basic number concepts and systems: *even*-versus-*odd*, *prime*-versus-*composite*, *integer*-versus-*fraction*, and so on. This suggests that the classificatory structure of the mathematical code is fundamentally oppositional (Danesi 2008). And this raises several fundamental questions about mathematics, such as: Are basic mathematical concepts necessarily oppositional? How is markedness assigned to mathematical concepts? Does markedness guide the learning and sequence of mathematical development, as some relevant research in psychology is beginning to show (for example, Cho, Procter 2007; Van der Schoot et al. 2009)? Clearly, in order to answer such questions a basic, rudimentary oppositional analysis of mathematical structure must be carried out beforehand.

The goal of this chapter is to revisit opposition theory in the light of its implications for the analysis of mathematical structure, extending it to encompass current approaches to mathematics from within the cognitive science field (for example, Lakoff, Nuñez 2000). Arguably, such a revisitation will make it possible to formulate specific questions about the nature of mathematical models and of the mathematical modeling system itself. Fermat proposed that the goal of mathematics was to explore the structure of the arithmetical forms and operations that make up its foundations. In his *Disquisitiones Arithmeticae* (1801), Gauss showed how number patterns form the core of mathematical structure. In both, the concept of opposition is implicit, since they look at binary forms such as *positive*-versus-*negative* and *real*-versus-*imaginary* as foundational structures in their discipline.

2.2. Opposition theory: An overview

It is instructive, first, to revisit opposition theory within semiotics, which actually traces its philosophical roots to Aristotle's logical dualism (Bogoslovsky 1928; Hjelmslev 1939; Babin 1940; Benveniste 1946; Bochénski 1961; Anfindsen 2006). The Aristotelian philosophical system was radicalized by Cartesian philosophy in the sixteenth century. But the Cartesian view was, and continues to be, more of an adaptation than a continuation of ancient philosophy, which actually aimed to understand the relation between matter and spirit, not their independence. As a psychological construct, dualism was implicit in the work of the early nineteenth-century psychologists Wilhelm Wundt (1880) and Edward B. Titchener (1910), and then in Saussure (1916) who recast it as the principle of *différence*. Saussure was the first to use the notion of binary opposition to explain how speakers are capable of extracting meaningful cues from the chain of speech. Starting in the late 1920s, the Prague School linguists (Jakobson et al. 1928; Jakobson 1932, 1936, 1939, 1968; Trubetzkoy 1936, 1939, 1968, 1975; Pos 1938, 1964) devised the first scientific theory of opposition, which they used to carry out extensive analyses of specific natural languages, thus founding structuralism as a mainstream school of analysis within linguistics, psychology, anthropology, and semiotics (Wallon 1945; Jakobson et al. 1952; Parsons, Bales 1955; Godel 1957; Lévi-Strauss 1958, 1971; Blanché 1966; Belardi 1970; Ivanov 1974; Needham 1973; Fox 1974, 1975; Lorrain 1975; Jakobson, Waugh 1979).

The first in-depth study of opposition as a psychological process was Charles K. Ogden's treatise, *Opposition: A Linguistic and Psychological Analysis* (1932), in which he expanded upon several key ideas he had discussed previously in 1923 in collaboration with I. A. Richards in *The Meaning of Meaning.* Ogden argued that a small set of binary oppositions, such as *right*-versus-*left* and *yes*-versus-*no*, were universal, since they appear across cultures and across time. Other (non-oppositional) concepts showed *gradience*, as Dwight Bolinger (1968) subsequently called it, and these could be located between the two poles of the universal binary oppositions. Gradient concepts are concepts that cannot be put into a binary relation. For instance, *white*-versus-*black* is a universal binary opposition, since color terms for these two are found throughout the world, but all other color concepts (*green, red*, etc.) are gradient ones that can be located between the *white* and *black* poles of the universal opposition, a fact that has both referential and conceptual resonance, since gradient colors are distributed on the light spectrum, while *white* and *black* are not:

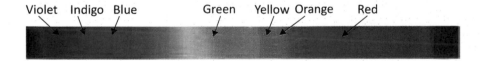

Gradient color concepts are thus culture-specific ones, varying from language to language, and devised according to need. The number that can be invented is limitless, but there is only one polar opposition, *white*-versus-*black*, which is often labeled more generically as a *clear*-versus-*dark* opposition (*chiaroscuro* in Italian). This opposition is actually purely conceptual since there is no real white or black color on the spectrum of light. *Black* is absence of color, and when seen together at the same time, the colors appear as white light. However, when light passes through a prism, the different colors separate and can be seen. This suggests that universal polar oppositions may have a basis in physical reality and that gradient ones are adaptive concepts emerging in specific semiospheres.

A similar story can be devised to explain the *day*-versus-*night* opposition, with these two constituting polar concepts, with related concepts such as *twilight, dawn, noon,* and *afternoon* forming gradient ones on that polar scale. In more technical terms, Ogden had discovered that polar concepts have paradigmatic structure, while gradient ones, being "distributed concepts" on oppositional scales, have syntagmatic structure. Ogden found, however, that oppositional structure can also be culture-specific. An opposition such as *town*-versus-*country* was a culturally-based one. In essence, a small set of polar concepts is universal, while others are culture-specific, as are all gradient ones.

Mathematical concepts also show a polar-versus-gradient structure. For example, the *whole*-versus-*fractional* opposition is found across cultures and across times, being expressed in various symbolic ways. But oppositions such as *rational*-versus-*irrational* are not found universally. These are discovered concepts, as we saw, and those form a gradient system. Within that system, however, concepts may bunch into oppositional pairs, creating new oppositional classes.

The technical framework for using opposition theory in the study of language was developed by the Prague School and extended by the Copenhagen School (Hjelmslev 1939, 1959). It became a mainstay in linguistics generally, and was even adopted by generativists to carry out phonological analyses (Chomsky, Halle 1968; McCarthy 2001; Haspelmath 2006). Starting in the 1980s, the framework was extended to encompass the study of discourse (Barbaresi 1988; Mettinger 1994; Mel'čuk 2001; Elšik, Matras 2006). Within semiotics, the spread of Peircean (1931–1958) theory in the late 1960s has led to a de-emphasis on the use of opposition theory to examine semiosis generally. However, this does not imply that Peircean theory was antithetical to Prague School structuralism. On the contrary, as can easily be extrapolated from current research on markedness and iconicity in language, Peircean and structural semiotics are highly compatible, complementary, versions of semiosis (for relevant research, see Andersen, H. 1989, 2001, 2008; Tomic 1989). Completing the historical picture are the insights of the Tartu School semioticians under the leadership of Juri Lotman (1991) who utilized opposition theory to show how different cultural codes are interconnected to each other in oppositional ways (see Andrews 2003; Lepik 2008; Semenenko 2012).

The work of the Prague School scholars laid the groundwork for all subsequent uses and extensions of opposition theory. Perhaps their greatest discovery was that oppositions are not *de natura* binary. There are in fact different types, levels, and manifestations of oppositional structure (Pos 1938; Jakobson 1939; Trubetzkoy 1939; Martinet 1960). One particularly crucial insight was that the signal in a phonemic opposition varied according to which phonemes were involved. For example, in the opposition /p/-versus-/b/ *(pin-versus-bin)* the relevant feature is *voice*. Both phonemes are occlusive and biliabial. What keeps them distinct is the fact the /p/ is *voiceless* and /b/ *voiced*. So, /p/ is marked with the feature [+voice] and /b/ as [-voice]. That single phonic feature is the trigger in the opposition. In a way, the scholars had split the phonemic atom, so to speak, showing what feature in the phoneme was the crucial one in determining an opposition. The same phoneme, however, was kept distinct by another feature when contrasted to another phoneme in an opposition. So, the feature that distinguishes the /b/ from /m/ in pairs such as *bad*-versus-*mad* is not [voice], since both phonemes are voiced and biliabial. In this case it is [nasal]: [+nasal] in /m/ but [-nasal] in /b/. The opposition-signaling feature came to be called a *distinctive feature.* And this led to the development of distinctive-feature analysis that is used to this day in phonology.

Work on distinctive features led, in turn, to a reconsideration of the nature of opposition and to a broadened typology of oppositional structures (Trubetzkoy 1939). In the area of phonology, the following oppositional features were found to apply:

- *multidimensional,* in which the distinctive features that are common to both phonemes also occur in other phonemes;
- *one-dimensional,* in which the features common to both phonemes do not occur in other phonemes;
- *isolated,* in which the features that occur in specific phonemic oppositions, occur nowhere else in the phonemic system;
- *proportional,* in which the features constituting certain phonemic oppositions are repeated in other phoneme pairs;
- *privative,* in which phonemes are distinguished by only one feature;
- *gradual,* which involves varying degrees of a feature in oppositional systems;
- *equipollent,* in which phoneme pairs are distinguished by several features.

The analysis of oppositions was soon extended to other levels of language and of meaning systems. A more generic subdivision of oppositions that was implicit in the work of the Prague School linguists can be labeled *form-based* and *conceptual.* Phonemic oppositions, for example, are form-based ones, because they involve distinctive physical cues in words, whereas oppositions such as *right*-versus-*left* and *day*-versus-*night* are conceptual, because they are based on differences in meaning. Applying distinctive-feature analysis to form-based oppositions was a straightforward and largely unproblematic procedure. How-

ever, problems emerged when such analysis was applied to conceptual oppositions. Pairs such as *man*-versus-*woman, boy*-versus-*girl*, for instance, could
be easily distinguished in terms of features such as [±human], [±gender], and
[±adulthood] called *semes* (Hjelmslev 1959; Coseriu 1973; Pottier 1974). However, if one compares sets of concepts that cut across categories how would one
distinguish between, say, *heifer*-versus-*mare*? Which distinctive feature trig-
gers the opposition, [±bovine] or [±equine]? There is really no way to establish
which one is conceptually the trigger when comparing cross-sets, as they can
be called. Moreover, a vast array of such features would be needed to make
conceptually-relevant distinctions in the first place (Schooneveld 1978), ren-
dering the whole analytical exercise artificial and highly circular. The alterna-
tive would be to decide upon a small axiomatic set of semes that cut across
languages and are thus deemed to be universal. Research on identifying such
a set is ongoing, but it has yet to yield a set of features that can truly explain all
kinds of conceptual oppositions (see, for example, the insightful work of Wierzbicka 1996, 1997, 1999, 2003). Unlike phonological and grammatical systems,
which are closed form-based systems, conceptual systems are open-ended and
thus difficult to pin down to a finite set of distinctive features.

Another problem with applying oppositional analysis to conceptual systems
concerns the nature of gradience. Abstract concepts such as *fatherhood*-versus-
motherhood, hope-versus-*despair*, for instance, are particularly high in connotative meaning. How would one go about determining this type of meaning
gradience? One answer was put forward in 1957 by the psychologists Charles E.
Osgood, George J. Suci, and Percy H. Tannenbaum with a technique that they
called the *semantic differential*. By using binary polar concepts such as *young*-
versus-*old, good*-versus-*bad*, etc. and asking subjects to rate a concept such as
fatherhood or *hope* on seven-point scales, in terms of the given polar concepts,
the technique would purportedly generate a statistically-valid connotative pro-
file of culture-specific gradience. The number seven was chosen because the
year before George Miller (1956) had shown that the ability to process bits of
information was limited to between 5 and 9 equally-weighted choices. Thus,
results near a polar end of the scales (say, 1.4 or 6.4) would indicate high or
low connotative content; results near the middle of the scales would indicate
neutrality and, thus, equipollence in the gradience. Although the semantic
differential seemed to be very promising as a technique for fleshing out the
connotative meanings attached to concepts, it was critiqued as being artificial,
since the polar concepts chosen as end-points on scales are predetermined by
the researchers and thus end up guiding subject ratings (constituting a kind of
unwitting Hawthorne Effect). In effect, the semantic differential produces what
analysts unconsciously believe it will produce. Yet in defense of the technique
one could say that even if the polar scales are determined in advance, the results
obtained may not be what was anticipated and thus the technique would end up
being a legitimate form of randomized experimentation.

Another discovery of the Prague School scholars was that oppositions were not limited to being binary. For example, the tense system of English has a basic ternary oppositional structure – *present*-versus-*past*-versus-*future*. In fact, they found that oppositions can be binary, ternary, four-part, graduated, or cohesive (set-based). The type of opposition that applies in a particular situation depends on the system under investigation (language, kinship, etc.). Anthropologist Claude Lévi-Strauss (1958), for example, showed that pairs of oppositions often cohered into sets forming recognizable units within specific cultural codes. In analyzing kinship systems, Lévi-Strauss found that the elementary unit of kinship was made up of a set of four binary oppositions: *brother*-versus-*sister*, *husband*-versus-*wife*, *father*-versus-*son*, and *mother's brother*-versus-*sister's son*. These then formed subsequent oppositions with others down the kinship hierarchy. In a similar train of thought, a decade later Algirdas J. Greimas (1966, 1970, 1987) introduced the notion of the semiotic square – a model of opposition involving two sets of concepts forming a square arrangement. Given a sign such as the adjective *rich*, Greimas claimed that we determine its overall meaning by opposing it not only to its contrary *poor*, as in binary oppositional analysis, but also to its contradictory *not rich* and to the contradictory of its contrary, *not poor*. This makes logical sense, of course, because one can be *not poor* and still not be *rich*. This type of analysis allows us to use contradictories such as *white*-versus-*non-white* and link them to contrary terms such as *white*-versus-*black* in a gradient fashion. In effect, by extending opposition theory, it was soon hypothesized that language, in the Saussurean sense of *langue*, had an "evaluative superstructure" that was oppositional in its overall makeup and design, but that the oppositions were multi-dimensional at different levels and for different purposes.

Of particular interest to the study of mathematical oppositions is Saussure's (1916: 251–258) notion of *valeur* (*value*) as the meaning that is generated about by the opposition. Thus, a minimal difference in sound, a minimal difference in tone, a minimal difference in orientation, and so on produces a *valeur*. The meaning we then assign to the *valeur* is what Saussure called their *substance*. Rather than carrying intrinsic meaning, Saussure argued, signs had *valeur* in differential relation to other signs or sign elements. To determine the *value* of an American quarter, for instance, one must know that the coin can be exchanged for a certain quantity (a *substance)* of something different and that its value can be compared with another value in the same system, for example, with two dimes and one nickel (Malmberg 1976). Counting with fingers, or with substitutive signs such as pebbles and knots, is an instinctive semiotic act that brings out the validity of *valeur* as an innate trait of human perception. In effect, people across the world recognize something as a number – as having contrastive numerical value – through its differential properties with other numbers. As discussed in the previous chapter, the constituent digits in decimal numerals, for instance, take on specific values not only in terms of their actual physical shapes, but also in terms of the positions they occupy in the numeral. Thus, we read the values of 7 and 3 not only in terms of their differential forms (7 has a different shape than 2 or in spoken language a different phonetics), but also in terms of their *valeur* in the position they occupy in a numeral: 73-versus-37.

2.3. Markedness

Related to the notion of opposition is that of markedness. Early on, it became obvious that in polar oppositions one of the two poles was perceived to be more "fundamental", and the other a kind of counterpart or derivative. This could often be interpreted as ensuing from a more fundamental opposition (a kind of meta-opposition): *presence*-versus-*absence*. In the case of an opposition such as *day*-versus-*night* it could be argued that *day* is the more fundamental concept – the default one – and that *night* is a derivative one, because in most cultures one would conceptualize *night* as "absence of day", and not the other way around – *day* as "absence of night".

The concept *day* was called "unmarked" and *night* "marked". At first, many polar oppositions could be related to the *presence*-versus-*absence* meta-opposition. Other factors were then discovered at the various levels of language as influencing markedness. For example, it was found that in a phonemic opposition, the marked phoneme tends to be more constrained in occurrence than the unmarked one (Morris 1938; Chomsky, Halle 1968; Hertz 1973; Trubetzkoy 1975; Jakobson, Waugh 1979; Waugh 1979, 1982; Tiersma 1982; Eckman et al. 1983). But at the conceptual level, the meta-oppositional structure is often subsumed. For instance, when an opposition such as *tall*-versus-*short* comes up in a speech situation, we typically ask "How *tall* are you?" not "How *short* are you?" because, unless there is a specific reason to do otherwise, we assume *tallness* to be the unmarked pole. This shows that the actual realization of the meta-opposition in a specific language is governed by contextual conditions. So, even an opposition such as the *day*-versus-*night* one can have its markedness structure reversed if, say, the people spoke a language in an area of the world where daylight was scarce at certain times of the year and thus work activities were conducted largely at nighttime – a situation that is not uncommon among people living near the earth's poles. So, in such languages, the marked pole would differ from what it is in other languages. It all depends on the physiological, psychological, historical, and contextual phenomena and situations that apply to human groups. This is essentially the point made by Benjamin Lee Whorf (2012: 182) in discussing not oppositional structure, but language categories in general:

If a race of people had the physiological defect of being able to see only the color blue, they would hardly be able to formulate the rule that they saw only blue. The term blue would convey no meaning to them, their language would lack color terms, and their words denoting their various sensations of blue would answer to, and translate, our words light, dark, white, black, and so on, not our word blue. In order to formulate the rule or norm of seeing only blue, they would need exceptional moments in which they saw other colors. The phenomenon of gravitation forms a rule without exceptions; needless to say, the untutored person is utterly unaware of any law of gravitation, for it would never enter his or her head to conceive of a universe in which bodies behaved otherwise than they do at the earth's surface. Like the color blue with our hypothetical race, the law of gravitation is a part of the untutored individual's background not something he or she isolates from that background. The law could not be formulated until bodies that always fell were seen in terms of a wider astronomical world in which bodies moved in orbits or went this way and that.

In the case of equipollent oppositions no markedness relation can be established as universal. This is the case in oppositions such as *town*-versus-*country* or *give*-versus-*receive*. Only the context in which they are used will determine how markedness is assigned to the poles. Moreover, it was found that markedness relations vary according to level. For instance, *grape* is less marked than *grapes* on the morphological level, since the singular form is typically the unmarked one on this level. However, on the semantic and discourse levels the singular *grape* is the marked one since the plural form *grapes* is referentially more common and thus unmarked. Research has also shown that there is a relation between markedness in language and social structure. In societies (or communities) where the masculine gender is the unmarked form, it is the men who tend to be in charge, while in societies (or communities) where the feminine gender is the unmarked form, the women are the ones who are typically in charge (Alpher 1987; King 1991). Markedness theory was thus found to be a diagnostic tool for unraveling unequal social relations and codes of power.

Work on opposition theory and markedness was attacked by a group of intellectuals starting as early as the late 1950s, but coming to prominence as a movement – *post-structuralism* – in the late 1960s and early 1970s. Michel Foucault (1972) and Jacques Derrida (1967) were leading figures in this attack, seeing structuralism as a product of logocentricism (Belsey 2002; Mitchell, Davidson 2007). Derrida came up with the notion of *deconstruction*, which he claimed would reveal the inbuilt biases built into texts that come from historical traditions and are imprinted in the words and oppositions used. By their very nature, sign systems are self-referential – signs refer to other signs, which refer to still other signs, and so on and so forth. For Derrida, words were empty structures, devoid of true meaning and, thus, could stand on their own for virtually anything. Deconstruction is rarely used today within semiotics proper as a main method of analysis, although it has left its residues in the ways in which texts are viewed. In the end, the deconstruction movement envisioned by Derrida was, arguably, nothing more than an overreaction to structuralism in its most radical forms. In effect, he attacked the very diagnostic tool that allowed him to flesh out problems in social systems in the first place. In polar concepts such as *day*-versus-*night* it is easy to accept *day* as the unmarked form and *night* as its marked counterpart, for various biological and psychocultural reasons. Problems emerge, as Derrida and Foucault correctly intimated, with oppositions such as *male*-versus-*female* and *self*-versus-*other*. But the very fact that such pairs were identified as problematic by structuralist theory in the first place is something that the post-structuralists missed or conveniently ignored. They ended up blaming the messenger for the message, constituting a classic example of *post hoc propter hoc* reasoning.

Post-structuralism never really affected work in Peircean semiotics or in the Tartu School of semiotics (Lotman 1991; Andrews 2003; Lepik 2008). Such work has shown, actually, how the meanings in a culture are interconnected in oppositional ways. As an example, consider the *right*-versus-*left* one (Needham 1973; Hertz 1973; Danesi 2007). This is derived, anatomically, from the fact that we have a left hand (and foot, leg, ear, and eye) and a right one. Now, this ana-tomical duality has been encoded in an opposition that carries a markedness criterion along with it – *right* is unmarked and *left* is marked in most societies. This very opposition can be seen to intersect with other oppositions – *right* is associated with *good, light,* etc. and *left* with *evil, dark,* etc. Cultures are thus defined as systems of interconnected codes based on intertwining opposition-al structures. Interestingly, basic oppositions seem to cut across cultures. As Ogden had already indicated, these seem to be wired into the human brain, including *masculine*-versus-*feminine, light*-versus-*dark, good*-versus-*evil, self*-versus-*other, subject*-versus-*object, sacred*-versus-*profane, body*-versus-*mind, nature*-versus-*culture, beginning*-versus-*end, love*-versus-*hate, pleasure*-versus-*pain, existence*-versus-*nothingness, left*-versus-*right, something*-versus-*nothing,* among others.

As an aside, it is instructive to note that markedness theory corresponds nicely with what has come to be called *prototype theory* in psychology, which traces its origin to 1958, when the psychologist Roger Brown (1958a) argued that children use what they know in concrete terms in order to refer to something in general. For example, Brown found that the word *dime* was used by most of his child subjects to refer both to the general concept of *money* and to specific types of dimes (such as a *2013 dime*). Brown referred to *dime* as a prototypical concept, to *money* as a superordinate concept, and to a *2013 dime* as a sub-ordinate concept. The follow-up research on prototypicality theory has largely confirmed Brown's claim that children learn abstract concepts at the prototypi-cal level and that, in general, prototypicality guides conceptual development and the grasp of concepts generally (Rosch 1973a, 1973b, 1975a, 1975b, 1981;

Rosch, Mervis 1975; Smith 1988; Taylor 1995). This finding meshes well with the work in markedness theory. Essentially prototypes are unmarked forms, whereas subcategories are essentially gradient concepts. Certain grammatical and lexical categories are perceived as more typical than others to speakers of a language. As a consequence, when a marked form does occur, it carries specific information along with it. It is also the form that is learned subsequently to unmarked ones as Jakobson (1942) showed early on.

We are, of course, aware that claiming certain forms as universal and others as culture-specific always entails a risk. A probability factor is thus involved. If something occurs frequently and across many cultures, then it can be considered to have universal status. But one should always bear in mind that even the linguist or semiotician has a worldview that is a result of upbringing. Benjamin Lee Whorf puts it eloquently as follows (cited in Danesi, Maida-Nicol 2012: 192):

> When linguists became able to examine critically and scientifically a large number of languages of widely different patterns, their base of reference was expanded; they experienced an interruption of phenomena hitherto held universal, and a whole new order of significances came into their ken. It was found that the background linguistic system (in other words, the grammar) of each language is not merely a reproducing instrument for voicing ideas but rather is itself the shaper of ideas, the program and guide for people's mental activity, for their analysis of impressions, for their synthesis of their mental stock in trade. Formulation of ideas is not an independent process, strictly rational in the old sense, but is part of a particular grammar and differs, from slightly to greatly, among different grammars. We dissect nature along lines laid down by our native languages. The categories and types that we isolate from the world of phenomena we do not find there because they stare every observer in the face; on the contrary, the world is presented in a kaleidoscopic flux of impressions which has to be organized by our minds – and this means largely by the linguistic systems in our minds. We cut nature up, organize it into concepts, and ascribe significances as we do, largely because we are parties to an agreement to organize it in this way – an agreement that holds throughout our speech community and is codified in the patterns of our language. The agreement is, of course, an implicit and unstated one, but its terms are absolutely obligatory; we cannot talk at all except by subscribing to the organization and classification of data which the agreement decrees.

As Whorf's statement implies, reality may be beyond probing and notions such as opposition are really nothing more than human fabrications devised to understand the world on human terms. Nevertheless, since mathematics is a human invention the semiotic categories and procedures described here apply to it. We will return to the topic of modeling versus reality in the final chapter.

2.4. Extending opposition theory

An approach to the study of the relation between language, culture, and cognition, known as conceptual metaphor theory (CMT), came to the forefront in the 1980s, after the publication of Lakoff and Johnson's landmark book, *Metaphors We Live By* (1980). CMT has become a major intellectual force within several interrelated fields, from linguistics to psychology and semiotics (Langacker 1987, 1990, 1999; Gibbs 1994; Gibbs, Colston 2012; Lakoff, Johnson 1999; Fauconnier, Turner 2002; Dirven, Verspoor 2004; Danesi 2004a; Geeraerts 2006; Müller 2008). Without going into specific details here, since CMT will be discussed in more detail subsequently in Chapter 4, suffice it to say that it has documented a critical fact – that there is a close relation between concepts and language structure, and that connective modeling (as it was called in the previous chapter) is a major process in the formation of all kinds of abstract concepts.

The underlying psychological mechanisms linking the various conceptual domains are called *image schemata* (Lakoff, Johnson 1980, 1999; Lakoff 1987; Johnson 1987). Upon closer scrutiny, these are essentially recast polar oppositions – *up*-versus-*down*, *back*-versus-*front*, *near*-versus-*far*, *full*-versus-*empty*, *balance*-versus-*unbalance*, etc. It can thus be argued that the whole apparatus of CMT is based on the structuralist notion of opposition. For this reason, the whole movement, as I have argued elsewhere (Danesi 2007), can really be considered to be a derivative and expansion of basic Prague School structuralism.

Lakoff and Núñez (2000) applied CMT theory for the first time in 2000 to explain the nature of mathematics. Essentially, they showed that mathematical notions and techniques such as proofs are interconnected through a process called *blending* (as already discussed in the previous chapter). This entails taking concepts in one domain and fusing them with those in another to produce new ones or to simply understand existing ones. Changing the blends leads to changes in mathematical structure and to its development. This line of argument coincides with the one put forward in the previous chapter, namely that discovery in mathematics results from insights produced by existing models. What is lacking, of course, in the Lakoff and Núñez (2000) explanatory framework is the consideration of metaphorical blends as producing models which Sebeok and Danesi (2000) called metaforms. Metaforms will be discussed in the fourth chapter. The difference is that a blend is a psychological construct, whereas a metaform is a semiotic one and thus one that involves understanding a model as something constructed, not as something resulting from a process.

Opposition theory is, as mentioned several times, fundamental in describing basic forms in mathematics. The *even*-versus-*odd* opposition, for example, occurs at the level of singularized forms, whereas the *linear*-versus-*quadratic* equation opposition occurs at the level of composite forms, and so on. In the as *even*-versus-*odd* opposition, the *even* one is the unmarked pole because odd numbers can be derived from the basic structure of even numbers, that is, even numbers show the structure 2n, so odd numbers are derived as follows: 2n + 1 (or 2n – 1). Cohesive oppositions, such as *rationals*-versus-*irrationals* allow for a higher-order level of modeling that pits systems of numbers and other mathematical forms against each other in order to glean meaningful values *(valeurs)* from them. Finally, connective modeling involves oppositions between types of metaforms such as proofs based on, say, a *deductive*-versus-*inductive* form.

The latter point requires some elaboration here. The ancient Greek philosophers had actually come to a truly insightful understanding of concept-formation over two and a half millennia ago. Basically, they claimed that many concepts are formed in one of two fundamental ways – by *induction* or by *deduction*. This became a basic opposition in their methodology of proving theorems. As is well known, the former involves reaching a general conclusion from observing a recurring pattern; the latter involves reasoning about the consistency or concurrence of a recurring pattern. Aware that other types of concepts existed (as those found in poetry, the arts, music, etc.), they argued that induction and deduction were particularly apt in explaining mathematical phenomena. Take, for example, the fact that the number of degrees in a triangle is 180°. One way to arrive at a practical knowledge of this fact is, simply, to measure the angles of hundreds, perhaps thousands, of triangles and then observe if a pattern emerges from the measurements. Assuming that the measuring devices are precise and that errors are not made, we are bound to come to the conclusion that the sum of the three angles adds consistently up to 180°. This "generalization-by-extrapolation" process is the sum and substance of inductive thinking. There is, of course, much more to inductive proof than this overly-reductive example implies. It is not the purpose here, however, to delve into its complexities. For the time being, it is sufficient to note that it results from a process of extrapolation.

Such demonstration is not, however, one hundred percent reliable, because one can never be sure that some triangle may not crop up whose angles add up to more or less than 180°. To be sure that 180° is the sum for *all* triangles one must use a deductive method of demonstration. This inheres in applying already-proved theorems or concepts to the case at hand. First, a triangle is constructed with a line parallel to the base going through its top vertex (A). The angles that form at the vertex are labeled with letters, as shown below:

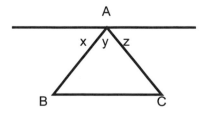

It is an established theorem of plane geometry that the angles on opposite sides of a transversal are equal. A transversal is a line that meets two parallel lines. In the diagram above, both AB and AC are transversals (in addition to being sides of the triangle). We use the theorem to label the equal angles with the same letters:

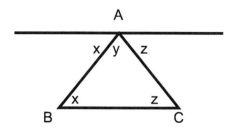

Now, we can use another established fact to show that the angles inside the triangle add up to 180° – namely, that a straight line is an angle of 180°. If we look at the parallel line going through A, we can see that the sum of the angles at A is x + y + z. Since these make up a straight line, we now can assert that x + y + z = 180°. Next, we look at the angles within the triangle and notice that the sum of these, too, add up to x + y + z. Since we know that this sum is equal to 180°, we have, in effect, proved that the sum of the angles in the triangle is 180°. Since the triangle chosen was a general one, because x, y, and z can take on any value we so desire (less than 180° of course), we have proved the pattern true for *all* triangles. This "generalization-by-demonstration" process is the sum and substance of deductive thinking.

Induction and deduction do, clearly, play a significant role in proof. But they hardly explain how the ideas, or more correctly hunches, behind proofs come about. It was of course Charles Peirce, himself a mathematician, who emphasized that many, if not most, of our insights are formed by a type of inferential process that he called *abduction*. He described it as follows:

> The abductive suggestion comes to us like a flash. It is an act of *insight*, although of extremely fallible insight. It is true that the different elements of the hypothesis were in our minds before; but it is the idea of putting together what we had never before dreamed of putting together which flashes the new suggestion before our contemplation. (CP 5.180)

The actual realization of proofs may show deductive or inductive form, but its original conceptualization – the way it is envisioned – is an abduction. The former two are useful as conceptual strategies for consolidating the hunches made by insight thinking: that is, they allow for mathematical discoveries to be organized into systems of mathematical knowledge. Once an insight is attained, it becomes useful to "routinize" it, so that a host of related problems can be solved as a matter of course, with little time-consuming mental effort. Such routinization is a memory-preserving and time-saving strategy. It is the rationale behind all organized knowledge systems. Such systems produce modeling strategies called *algorithms* – routinized procedures – for solving problems that would otherwise require insight thinking to be used over and over again.

2.5. Opposition in mathematical modeling

As we have seen above, applying opposition theory to mathematics turns out to be a rather straightforward procedure. Consider, for example, a basic opposition in the integer system, namely the one between *positive* and *negative* numbers. The positive ones are represented with the + sign in front:

$$\{+1, +2, +3, +4, \ldots\}$$

However, it is standard practice not to use this sign, unless it is needed for some specific purpose. This is so because such numbers are considered to be the unmarked ones of the opposition. For this reason they are not marked with the + sign unless it is required for some specific reason:

$$\{+1, +2, +3, +4, \ldots\} = \{1, 2, 3, 4, \ldots\}$$

On the other hand, negative integers, being marked forms, are always identified with the minus sign – in front:

$$\{-1, -2, -3, -4, \ldots\}$$

The complete set of integers (positive and negative) includes 0, which separates the positive and negative ones (as for example on a number line).

$$\{\ldots, -4, -3, -2, -1, 0, 1, 2, 3, 4, \ldots\}$$

Now, this opposition leads to the construction of a basic model of numeration and numerality, leading to implications and applications of various kinds. Negative numbers are used for a variety of practical reasons in daily life. For example, any temperature marking a point below the zero on a thermometer is indicated by a negative number. The temperature marking "5 below zero Centigrade" is represented as –5 C°. The *positive*-versus-*negative* opposition leads to a related number of concepts. One of these is the number line, itself a diagrammatic model of this opposition, in which the distance between any two numbers is shown as constant and, thus, the magnitude or absolute value (*valeur*) of each number equals its distance on the line from zero:

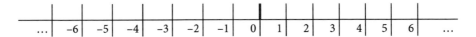

Obviously, one can keep on inserting numbers *ad infinitum* to the right or to the left. This simple diagram thus shows concretely that the set of all integers, positive and negative, is *infinite* (never-ending).

It is relevant to note that, historically, the concept of negative number took a while to surface in mathematical practice, as is consistent with the psychological theory of markedness. It probably crystallized first in China, since the opposition of *positive*-versus-*negative* is found in a 250 BCE text titled *Chui-chang swan-shu* (The Nine Chapters). During the seventh century negative numbers appear in the bookkeeping practices and astronomical calculations of the Indians. It was not until the sixteenth century that such numbers surfaced in Europe, appearing in Girolamo Cardano's works. It is not coincidental that the term *negative* comes from the Latin past participle of *negare* ("to deny"), perhaps because the existence of such numbers had been denied for so long, or because it implied philosophically the "denial" of the positive, so to speak.

The number line is a derivative model (discussed in the previous chapter), and an extremely useful one at that. It is a model of counting procedure. For example, adding 8 + 4 involves counting linearly in two steps: from 0 to 8 and then four more steps to 12:

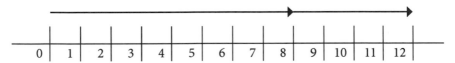

Addition notation is really shorthand for this kind of reasoning and modeling. Converting the above diagram into appropriate addition notation and layout, such as: 8 + 4 = 12 or 4 + 8 = 12 is part of algorithmic procedure, not conceptual knowledge. What these forms allow us to do is avoid repeating the kind of reasoning just described over and over each time we add. In other words, as representamina they make addition mechanical and routine.

Subtraction is the opposite of addition – the differential factor of the opposition being the direction of the counting process on the number line. For example, if we wish to subtract 5 from 9, we reverse the linear counting procedure, that is, we make it go in the opposite direction of addition. So, we count 5 "down" (to the left) from 9, reaching 4 on the number line:

Given the conceptual structure of multiplication as repeated addition, we can now model the operation appropriately using the number line. For instance, $2 \times 3 = 6$ can be shown as 2 added to itself three times to the right of 0 on that line, reaching the end point 6:

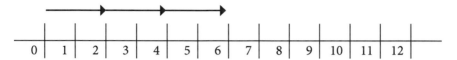

Notice that the same result would be obtained by adding 3 to itself twice, in conformity with the commutative law of multiplication: $2 \times 3 = 3 \times 2$. Multiplication notation is shorthand for this kind of reasoning. Division is repeated subtraction and the reverse of multiplication. The number line can thus be used again to model the repeated subtraction process. For example, $6 \div 3 = 2$ indicates that we must subtract 3 twice from 6. The "twice" is the end result, of course, of reaching zero and thus the answer.

The plus and minus signs are clearly signifiers of directionality, with the former representing right-hand movement and the latter left-hand movement on the number line. A positive number +n differs from its oppositional counterpart -n by being at the same point in distance from 0 but "on the other side". It forms a perfect indexical binary opposition. Adding and subtracting signed numbers can thus be modeled simply as directions to go left or right on the number line, according to the given sign. For example, adding (+2) and (+3), or simply 2 + 3, means go right two whole units from zero, and then three more to the right. The end point (or answer) is (+5):

Adding two negative numbers can be modeled easily as well. For example, (−3) + (−2) tells us to go left three units from 0 and then left two more. The end point (or answer) is (−5):

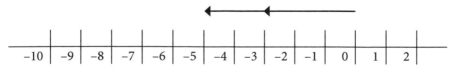

The sum (end point) of two negative numbers is also negative. To add numbers with different signs, the same type of reasoning applies. Consider (+5) + (−7). The first number tells us to go right five units from 0. The second tells us to change direction and go left seven units from there. The end point (or answer) is (−2):

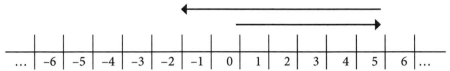

The sign of the number with the larger absolute value (−7) is also the sign of the answer (end point). And the reason for this is evident on the number line. Adding a positive and a negative number is really a subtraction problem. If the signs are the same, we add the numbers and give the sum the common sign (as shown above). If the signs are different, we subtract the number with the smaller absolute value from the number with larger absolute value, assigning to the result the sign of the larger one. If a negative sign precedes a negative number, we change its sign to positive.

Given the nature of multiplication as repeated addition, to multiply, say, (+2) × (+3), we go right two units from 0 three times in sequence. The end result is the point (+6):

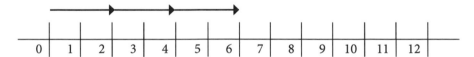

Now, if the signs of the two digits are different, then their product is negative: for example, (−3) × (+2) = (−6). This tells us that (−3) must go in the "same direction" (to the left) twice:

The basic reasoning used for the division of unsigned numbers applies to the division of signed numbers. For example, (+12) − (+4) = (+3) tells us that we have to subtract 4 units from (+12) three times:

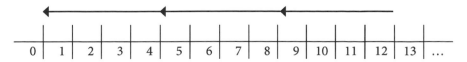

Knowledge of how to represent number oppositions (with formulas) is extremely important in mathematics. As discussed, even numbers are represented with 2n and odd numbers as 2n + 1. Here are some questions that the *2n-versus-2n + 1* opposition can allow us to answer.

(1) What is the result of adding two even integers?
(2) What is the result of adding two odd integers?
(3) What is the result of adding an even and an odd integer?
(4) What is the result of multiplying two even integers?
(5) What is the result of multiplying two odd integers?
(6) What is the result of multiplying an even and an odd integer?

The answer to (1) is even because $2n + 2n = 4n$ and $4n$ is itself even: $2(2n)$. The answer to (2) is also even because $(2n + 1) + (2n + 1) = 4n + 2$ and this is itself even: $2(2n + 1)$. The answer to (3) is odd because $2n + (2n + 1) = 4n + 1$ and this is itself odd: $(2 \times 2n) + 1$. The answer to (4) is even because $(2n)(2n) = 4n^2$ and this is itself even: $2(2n^2)$. The answer to (5) is odd because $(2n + 1)(2n + 1) = 4n^2 + 4n + 1$ and this is itself odd: $2(2n^2 + 2n) + 1$. Finally, the answer to (6) is even because $(2n)(2n + 1) = 4n^2 + 2n$ and this is itself even: $2(2n^2 + n)$. In all these problems the basic structure of even and odd numbers is examined and recast through substitutions and equivalences.

Another basic opposition that applies to the integers is that of *prime-versus-composite*. This opposition came into being because of the fact that it was noticed by Pythagoras that some numbers can be decomposed into smaller numbers, called factors, which make them up through multiplication:

$1 \times 12 = 12$
$2 \times 6 = 12$
$3 \times 4 = 12$

Some factors can be decomposed further. For example, the factor 4 in the product $3 \times 4 = 12$ above can itself be expressed as the product of $2 \times 2 = 4$. We can plug this in as follows:

$3 \times 4 = 3 \times (2 \times 2) = 3 \times 2 \times 2 = 12$

This is also the case for $2 \times 6 = 12$, which can be decomposed to the same set of three factors, because 6 is itself composed of the factors $3 \times 2 = 6$:

$2 \times 6 = 2 \times (3 \times 2) = 2 \times 3 \times 2 = 3 \times 2 \times 2 = 12$

These are the skeletal factors of 12, since 3 and 2 cannot be decomposed further. This type of analysis yields a fundamental insight into the structure of the integers and, thus, a central opposition within the integer code. The factors that cannot be decomposed further are called *prime numbers*; all the others are called *composite*. The definition of a prime is, in fact, any number that has no factors other than, of course, itself and the number 1 (which is a factor in every single number). For instance, the number 7 is prime because 7 and 1 are its only factors: $7 \times 1 = 7$. The first eight primes are:

$$\{2, 3, 5, 7, 11, 13, 17, 19\}$$

Numbers are either prime or composite. Since composites are more frequent, they are unmarked forms, whereas primes are much less frequent and are thus highly marked. Both types of numbers, however, are infinite. This *prime*-versus-*composite* opposition is one of the most fascinating of all numerical oppositions, having a long and fascinating history behind it. For example, as it turns out, each composite number is made up of one, and only one, distinct set of prime factors. This is called as the *Fundamental Theorem of Arithmetic*. It was formulated by Euclid (in Book IX of the *Elements*), and proved by the German mathematician Carl Friedrich Gauss (1777–1855) two millennia later. As a composite form of proof, the theorem is worth revisiting here. Given any composite number, the theorem states that it is decomposable into a unique set of prime factors. Take 24 as a case in point. The steps that can be used to reduce it to its skeletal prime structure unfold as follows:

(1) $24 = 12 \times 2$
(2) We notice that $12 = 6 \times 2$
(3) We plug this in (1) above: $24 = (6 \times 2) \times 2 = 6 \times 2 \times 2$
(4) We notice that $6 = 3 \times 2$
(5) And we plug this in (3) above: $24 = 6 \times 2 \times 2 = (3 \times 2) \times 2 \times 2 = $
$= 3 \times 2 \times 2 \times 2$

We have now uncovered the prime factors of 24. They are 2 and 3. Of course, the 2 occurs three times and the 3 once: $24 = 3 \times 2 \times 2 \times 2 = 3 \times 2^3$. Each of the prime factors in a composite number also divides evenly into it: 3 divides into 24 as does 2. This suggests a simple method for determining the prime factors of any number. Take 220 as a case in point. We can start by checking if the smallest prime number, 2, divides into the number evenly:

$$220 \div 2 = 110$$

We continue dividing by 2 until it is no longer possible to do so (evenly):

$$110 \div 2 = 55$$

The result of 55 cannot be divided evenly by 2. So, we go to the next smallest prime in the set of primes, 3. It does not divide evenly into 55 either. So, we go on to the smallest prime after 3, which is 5. It does divide into 55 evenly:

$$55 \div 5 = 11$$

The result, 11, is itself a prime number, bringing the division procedure to an end. The prime factors of 220 are, therefore: 2 (twice), 5, and 11:

$$220 = 2 \times 2 \times 5 \times 11 = 2^2 \times 5 \times 11$$

The amount of debate and research that the *prime-versus-composite* opposition has set off in the history of mathematics is astronomical. One of the first to use this opposition ingeniously was Euclid, who proved that the number of primes is infinite. The proof seems to contradict common sense, since primes appear to become scarce as the numbers grow larger: twenty-five percent of the numbers between 1 and 100, 17 percent of the numbers between 1 and 1000, and 7 percent of the numbers between 1 and 1,000,000 are primes. Euclid's proof is worth revisiting here.

To start off, Euclid noted that composite numbers are the products of prime factors. He then assumed that there may indeed be a finite set of primes, labeling them as follows (P = set of all primes).

$$P = \{p_1, p_2, p_3, \dots p_n\}$$

The symbol p_n stands for the last (largest) prime; each of the other symbols stand for a specific prime. So, $p_1 = 2$, $p_2 = 3$, and so on. Concretely, the set would look like this:

$$P = \{2, 3, 5, 7, 11, \dots p_n\}$$

Euclid then had one of the illuminations that come from considering pattern: What kind of number would result from multiplying all the primes in the set? Let's call this number N:

$$N = \{p_1 \times p_2 \times p_3 \times \dots \times p_n\}$$

Concretely, this would look like this:

$$N = \{2 \times 3 \times 5 \times \dots \times p_n\}$$

N would, of course, be a composite number because it can be factored into smaller prime factors – p_1, p_2, etc. Then, Euclid added 1 to this product: $(p_1 \times p_2 \times p_3 \times \ldots \times p_n) + 1$. Let's call the number produced in this way M:

$$M = \{p_1 \times p_2 \times p_3 \times \ldots \times p_n\} + 1.$$

Clearly, M is not decomposable, because when any of the prime factors available to us $\{p_1, p_2, p_3, \ldots p_n\}$ are divided into it, a remainder of 1 would always be left over. So, the number M is either: (1) a prime number that is not in P and obviously much greater than p_n, or (2) a composite number with a prime factor that, as just argued, cannot be found in the set $\{p_1, p_2, p_3, \ldots p_n\}$ and thus also greater than p_n. Either way, there must always be a prime number greater than p_n. In this way, Euclid showed that the primes never end.

Euclid's proof motivated mathematicians to come up with a formula for generating the prime numbers – but so far to no avail. One of the first attempts to create a formula or algorithm for the generation of primes was the so-called Sieve of Eratosthenes, invented in the 200s BCE by the Greek mathematician Eratosthenes of Cyrene (c. 276–195 BCE):

1	2	3	4	5	6	7	8	9	10
11	12	13	14	15	16	17	18	19	20
21	22	23	24	25	26	27	28	29	30
31	32	33	34	35	36	37	38	39	40
41	42	43	44	45	46	47	48	49	50
51	52	53	54	55	56	57	58	59	60
61	62	63	64	65	66	67	68	69	70
71	72	73	74	75	76	77	78	79	80
81	82	83	84	85	86	87	88	89	90
91	92	93	94	95	96	97	98	99	100

To flesh out the prime numbers, the following must be done:
- Cross out every second number after 2, a procedure that eliminates all numbers that can be divided evenly by 2, except for 2 itself:

    {2, 3, 4̶, 5, 6̶, 7, 8̶, 9, 1̶0̶, 11, 1̶2̶, 13, 1̶4̶, 15, 1̶6̶, 17, 1̶8̶....}

- Then cross out every third number after 3, a step that eliminates all the numbers that can be divided evenly by 3, except for 3 itself:

    {2, 3, 4̶, 5, 6̶, 7, 8̶, 9̶, 1̶0̶, 11, 1̶2̶, 13, 1̶4̶, 1̶5̶, 1̶6̶, 17, 1̶8̶....}

- Since 4 is already crossed out, move on to 5. Continue the same process. Any number that has not been crossed out is prime. The primes can be thought of as having passed through a sieve (strainer) that has caught all the composites.

Prime numbers are the building blocks of the whole architecture of arithmetic. In a letter to Euler in 1742, the mathematician Christian Goldbach conjectured that every even integer greater than 2 could be written as a sum of two primes:

$$4 = 2 + 2$$
$$6 = 3 + 3$$
$$8 = 5 + 3$$
$$10 = 7 + 3$$
$$12 = 7 + 5$$
$$14 = 11 + 3$$
$$16 = 11 + 5$$
$$18 = 11 + 7$$
$$\dots$$

No exception is known to Goldbach's Conjecture, as it has come to be known, but we have no valid proof of it either. Goldbach also conjectured that any number greater than 5 could be written as the sum of three primes:

$$6 = 2 + 2 + 2$$
$$7 = 2 + 2 + 3$$
$$8 = 2 + 3 + 3$$
$$9 = 3 + 3 + 3$$
$$10 = 2 + 3 + 5$$
$$11 = 3 + 3 + 5$$
$$\dots$$

To this day, no formula has ever been devised to generate primes and only primes. The sixteenth century mathematician Marin Mersenne claimed he had come up with one such formula: $2^n + 1$. The formula does indeed generate a high number of primes, known today as Mersenne numbers. But many exceptions have been found.

$$n = 2$$
$$2^n - 1$$
$$2^2 - 1 = 4 - 1 = 3 \text{ (prime)}$$

$$n = 3$$

$$2^n - 1$$
$$2^3 - 1 = 8 - 1 = 7 \text{ (prime)}$$

$$n = 4$$
$$2^n - 1$$
$$2^4 - 1 = 16 - 1 = 15 \text{ (composite)}$$
$$\ldots$$
$$n = 19$$
$$2^n - 1$$
$$2^{19} - 1 = 524{,}288 - 1 = 524{,}287 \text{ (prime)}$$
$$\ldots$$

Mersenne claimed that $2^{31} - 1$ was prime, and it was proved to be prime only much later in 1772; on the other hand, Mersenne also maintained that $2^{67} - 1$ and $2^{257} - 1$ were prime, but this was proved later to be wrong. By 1914, mathematicians had discovered a Mersenne number that was 29 digits long. Today, computers use the formula to keep searching for longer and longer primes. Called the Great Internet Mersenne Prime Search (GIMPS), and started in 1996, volunteers from across the globe continue to search for larger and larger Mersenne primes with GIMPS (at least at the time of the writing of this book).

The poles of certain binary oppositions (*addition*-versus-*subtraction, multiplication*-versus-*division*) are conceptualized as reversals of each other. In these, markedness is assigned accordingly. In the *addition*-versus-*subtraction* opposition, *addition* is the unmarked one and *subtraction* the marked one; similarly, in the *multiplication*-versus-*division* opposition, *multiplication* is the unmarked one and *division* the marked one. Now, when considering all four together, it can be seen that there are orders of oppositions involved. A basic opposition, such as *addition*-versus-*subtraction,* can be called a first-order opposition, and the derived opposition, *multiplication*-versus-*division,* can be called a second-order opposition (see also Beziau, Payette 2012). This four-part oppositional system is consistent with both the traditional teaching order of the four operations and the fact that the four operations are related to each in specific (derivative) ways.

The concept of order also allows us to grasp the connection between levels in mathematics. The primary order is the instinctive ability itself to count. Using numerical signs to stand for counting constitutes a secondary order. This is a representational order and can be called, simply, the order of numeration. It is the level at which counting concepts are semiotized into numerals. The third order is the arithmetical one. This is the level at which numerals are organized into a code of operations based on counting processes (adding, taking away, comparing, dividing, and so on). Finally, algebra constitutes a fourth-order language that has the capacity to generalize the features and patterns of arithmetic by simply representing them with signs called variables and extending the oppositional structure of numerals to variables. The emergence of algebraic competence is, more than likely, the end result of following such a "natural semiotic order", as it can be called. Simply put, algebra involves the semiotic ability to represent arithmetical structures and oppositions in general ways.

Fourth Order	Algebra = developing a metalanguage for arithmetic
Third Order	Arithmetic = encoding counting processes into rules and oppositions
Second Order	Numeration = using numeral signs to stand for counting units
First Order	Counting = the instinctive ability to separate referents into units that can be compared to each other in terms of quantity

Several fundamental questions emerge from the analysis put forward in this chapter. One is: as interesting as it is, does opposition theory really explain mathematics, language, or anything else, or is it nothing more than a figment of the fertile minds of linguists and semioticians? It was Jakobson (1942) who first dealt with this question empirically by documenting the stages of the child's linguistic development, verifying the fact that infrequent oppositions are among the last ones learned by children – a fact that cuts across learning systems (see, for example, Collins 1969; Barbaresi 1988; Park 2000; Mansouri 2000; Schuster 2001; Hatten 2004; Arranz 2005; Vijayakrishnan 2007). An important recent study by Van der Schoot et al. (2009) examined the opposition *more than*-versus-*less than* (the first being the unmarked term) in word problem solving in 10–12-year-old children differing initially in problem-solving skills. The researchers found that the less successful problem solvers utilized a successful strategy only when the primary term in a problem was the unmarked one. In another significant study, Cho and Proctor (2007) found that classifying numbers as odd or even with left-right keypresses was carried out more successfully with the mapping *even-right*-versus-*odd-left* than with the opposite mapping. Calling this a *markedness association of response codes* (MARC) *effect*, the researched attributed it to compatibility between the linguistic markedness of stimulus and response codes.

A second major question regards the notion of distinctive features: What kinds of distinctive features are involved in mathematical oppositions? What distinctive feature, for example, keeps odd numbers apart from even numbers?

Is the feature [divisibility by 2]? This kind of question has never been explored to the best of our knowledge. It might well be that distinctive-feature analysis does not apply to mathematics and, thus, that oppositions are highly conceptual. This is an area that needs to be explored further. A third major question, and the most important one, is: to what extent or in what way are forms such as the Pythagorean theorem oppositional? The Pythagorean theorem holds in two-dimensional space. It does not in other spaces, nor do any of the facts pertaining to planar triangles. As is well known, by applying the triangle form to a globe, the angles in the triangle will be greater than $180°$. In the figure below $a + b + g > 180°$.

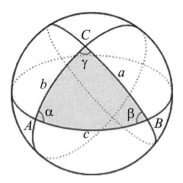

This leads to the fact that not all structure is oppositional, even though it may typify a large swath of mathematics, but other concepts, such as the Pythagorean theorem, non-Euclidean geometry (derived from considerations such as that above), and the theory of infinite sets may not. Let us consider the latter briefly here, and especially Cantor's discovery of different infinities, which would not have coalesced in his mind without the consideration of numbers as infinite points on a number line. In the sixteenth century Galileo noticed a pattern – if the counting numbers, {1, 2, 3, 4, …}, are compared, one-by-one, to the numbers in one of its subsets in sequence, such as the square numbers {1, 4, 9, 16, …}, something astounding happens:

1	2	3	4	5	6	7	8	9	10	11	12	13	14	…
↕	↕	↕	↕	↕	↕	↕	↕	↕	↕	↕	↕	↕	↕	…
1	4	9	16	25	36	49	64	81	100	121	144	169	196	…

No matter how far one continues along this layout of numbers, there will never be a gap between the top line and the bottom one. This suggested to Galileo that the "number" of elements in the set of all positive integers and the "number" of those in one of its proper subsets is the same. In the 1870s, this very same concept inspired Cantor's theory of infinite sets. Like Galileo, he showed that the set of counting numbers, also called cardinal numbers, can be matched against any of its subsets, such as the even numbers:

1	2	3	4	5	6	7	8	9	10	11	12	13	14	...
$\updownarrow$	$\updownarrow$	$\updownarrow$	$\updownarrow$	$\updownarrow$	$\updownarrow$	$\updownarrow$	$\updownarrow$	$\updownarrow$	$\updownarrow$	$\updownarrow$	$\updownarrow$	$\updownarrow$	$\updownarrow$	...
2	4	6	8	9	10	11	12	13	14	15	16	17	18	...

Cantor pointed out that such sets have the same "cardinality", or same number of elements, which he represented with the Hebrew letter $\aleph_0$ (aleph null):

$$\aleph_0 = \{1, 2, 3, 4, ...\}$$

Any infinite set that can be put in a one-to-one correspondence with $\aleph_0$ has the same cardinality (size). He called $\aleph_0$ a transfinite number, and then discovered something else that was even more mind-boggling – the transfinite numbers themselves constitute an infinite set $\{\aleph_0, \aleph_1, \aleph_2, \aleph_3, ...\}$, each one describing sets with different cardinalities! The set of real numbers, for example, has a greater cardinality than $\aleph_0$.

Cantor also showed that some infinities are larger than others. Here's how he did it. The set of fractions seems to be much larger than that of the positive integers, since on the number line there are infinitely more fractions than whole numbers, since infinities of fractions can be put between two whole numbers. But Cantor used an ingenious diagram that allowed him to list all fractions in order, making them countable. He simply listed all the fractions as shown below in a diagonal way: the first row lists the integers, the second row lists the "halves" of the integers above, the third row the "thirds," and so on. By following the lines around the diagonals of the array and deleting the numbers that are repeated the following picture emerges:

$$
\begin{array}{llllllll}
1/1 & 1/2 \rightarrow 1/3 & 1/4 \rightarrow 1/5 & 1/6 \rightarrow 1/7 & 1/8 \rightarrow \cdots \\
2/1 & 2/2 & 2/3 & 2/4 & 2/5 & 2/6 & 2/7 & 2/8 & \cdots \\
3/1 & 3/2 & 3/3 & 3/4 & 3/5 & 3/6 & 3/7 & 3/8 & \cdots \\
4/1 & 4/2 & 4/3 & 4/4 & 4/5 & 4/6 & 4/7 & 4/8 & \cdots \\
5/1 & 5/2 & 5/3 & 5/4 & 5/5 & 5/6 & 5/7 & 5/8 & \cdots \\
6/1 & 6/2 & 6/3 & 6/4 & 6/5 & 6/6 & 6/7 & 6/8 & \cdots \\
7/1 & 7/2 & 7/3 & 7/4 & 7/5 & 7/6 & 7/7 & 7/8 & \cdots \\
8/1 & 8/2 & 8/3 & 8/4 & 8/5 & 8/6 & 8/7 & 8/8 & \cdots \\
\vdots & \vdots & \vdots & \vdots & \vdots & \vdots & \vdots & \vdots & \ddots
\end{array}
$$

As can be seen, eliminating repetitions, this technique produces all the positive fractions:

1/1 (or 1), 2/1 (or 2), 1/2, 1/3, 3/1 (or 3), 4/1 (or 4), 1 1/2, 2/3, 1/4, 5/1 (or 5), …

The path indicated by the lines, therefore, allows us to set up a one-to-one correspondence between the integers and the array numbers. This shows clearly that there are as many fractions as there are whole numbers.

Integers	1	2	3	4	5	6	7	8	9	10	11	12	13	…
	↕	↕	↕	↕	↕	↕	↕	↕	↕	↕	↕	↕	↕	
Array	1/1	2/1	1/2	1/3	3/1	4/1	3/2	2/3	1/4	1/5	5/1	6/1	5/2	…

To list all the fractions (positive, negative, and zero) in order, all we have to do is alternate + and −, as before. This shows that the set of all fractions is countable.

Cantor also argued that the set of all real numbers is not countable. All we have to do is prove that the set of all numbers between 0 and 1 on the number line is not countable. Here's how he did it. Let us lay the numbers out in decimal form, labeling them as follows: $\{N_1, N_2,...\}$. There are so many possible numbers of the form p/q between 0 and 1 that we could not possibly put them in any order. So, the numbers given here are just a sampling:

$N_1 = .4225896\ldots$
$N_2 = .7166932\ldots$
$N_3 = .7796419\ldots$

$\ldots$

How could we possibly construct a number that is not on that list? Let's call it C. To create it, we do the following: (1) for its first digit after the decimal point we choose a number that is greater by one than the first digit in the first place of N_1; (2) for its second digit choose a number that is greater by one than the second number in the second place of N_2; (3) for its third digit choose a number that is greater by one than the third number in the third place of N_3; (4) and so on:

$N_1 = \underline{4}225896\ldots$
The constructed number, C, would start with 5 rather than 4 after the decimal:
$C = .5\ldots$

$N_2 = .7\underline{1}66932\ldots$
The constructed number would have 2 rather than 1:
$C = .52\ldots$

$N_3 = .77\underline{9}6419\ldots$
The constructed number would have 0 rather than 9:
$C = .520\ldots$

etc.

Now, the number $C = .520\ldots$ is different from N_1, N_2, N_3, ... because its first digit is different from the first digit in N_1; its second digit is different from the second digit in N_2; its third digit is different from the third digit in N_3, and so on *ad infinitum*. We have in fact just constructed a number that appears nowhere in the list.

These demonstrations are staggering both for the incredible insights they produce on the nature of infinity and for their simplicity. The point to be made here is that although they do not seem to be implanted on oppositional thinking, they could not have been conceptualized in the first place without the existence of oppositional categories in mathematics. In discussing the applications of semiotics to the study of mathematics, even Cantor could not believe the results he obtained with his simple models, no matter what their real-world implications might be or not be. One of his demonstrations was truly astonishing in defying common sense and conventional logic. Traditionally a square is seen to acquire its form from the idea that it consists of lines of unit length piled one on top of the other. If one of these lines contains a number of points that can be marked, then a square, consisting of numerous lines, will presumably have many more points within it. But Cantor showed that this is not so. He did this by choosing a point on the line as a decimal, for instance, P= 0.19762543… He then constructed a corresponding point in the square with coordinates (x, y) by means of alternately allocating the digits to the coordinates, so that we would have $x = 0.1724$ and y = 0.9653… In reverse, if we are given a point in the square (x, y) we can interlace the digits of x and y to form one decimal expression corresponding to a point on the line:

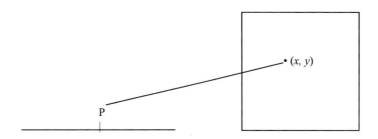

In other words, one can set up a one-to-one correspondence with all points on the line, P_n, with every point (x_n, y_n) in the square. This is truly an incredible proof. Cantor was even able to show that the same holds between a line and a three-dimensional figure such as a cube. As Crilly (2011: 119) writes, Cantor himself could not believe what he had proved:

> What Cantor had done flew in the face of intuition, even his own. He wrote to Richard Dedekind in 1877, remarking: "I see it, but I don't believe it".

The persuasive power of Cantor's proofs is inescapable, but it is not a kind of cognitive process that can easily be described. It is a form of insight thinking that comes from using diagrams and other visual prompts. As Fauconnier and Turner (2002) suggest, proofs such as those by Cantor are blends – the result of connective modeling. Once completed, a blend is available for incorporation into additional blends. To cite Turner (2005):[1]

> As long as mathematical conceptions are based in small stories at human scale, that is, fitting the kinds of scenes for which human cognition is evolved, mathematics can seem straightforward, even natural. The same is true of physics. If mathematics and physics stayed within these familiar story worlds, they might as disciplines have the cultural status of something like carpentry: very complicated and clever, and useful, too, but fitting hu-

---

[1] Available as Turner, Mark 2005. Mathematics and Narrative. thalesandfriends.org/en/papers/pdf/turner paper.pdf.

man understanding. The problem comes when mathematical work runs up against structures that do not fit our basic stories. In that case, the way we think begins to fail to grasp the mathematical structures. The mathematician is someone who is trained to use conceptual blending to achieve new blends that bring what is not at human scale, not natural for human stories, back into human scale, so it can be grasped.

To conclude this chapter, it is correct to say that as one of the most important achievements of the Prague School, opposition theory continues not only to have validity as an investigative tool into the structure of human codes, such as mathematics, but also significant implications for uniting various approaches in the human sciences to the study of cognition and culture. It is one of those notions that has always been implicit in human affairs, but which needed articulation in a concrete scientific way. But even the notion of opposition really does not penetrate the enigma of mathematics and the various questions that it elicits in this domain of knowledge, as just discussed. It may only scratch the surface, as the work by Cantor, Gödel, and others implies. Also, such fields as topology and fractal geometry have no foreseeable oppositional structure, at least in the traditional sense of that term. Opposition theory also skirts what is perhaps the most important question of all: Why do artifactual mathematical models manifest themselves serendipitously in natural forms (Ghyka 1977; Schneider 1994; Stewart 1995; Clawson 1999; Ball 2003; Adam 2004)? Why this is so remains one of the greatest mysteries of all times. It is a topic that will be explored further in the final chapter of this book.

3. Diagrammatic modeling in mathematics

3.1. Introduction

Let us return again to the Pythagorean theorem and its many proofs. The typical method for demonstrating the truth of the theorem is, as discussed previously, diagrammatic, revealing that mathematics has always been a modeling system, using primarily diagrammatic models to conduct its exploratory and theoretical affairs (Barker-Plummer, Bailin 1997, 2001; Kulpa 2004; Stjernfelt 2007). One of the simplest of all diagrams showing the Pythagorean relation between the sides of a right triangle is the following:

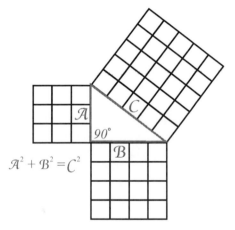

This diagram shows several things at once. First, it is pictorial evidence that the squares on the sides of the triangle (segmented into smaller unit squares) add up to the number of such units in the large square on the hypotenuse. Second, it shows the relation between the equation and the actual diagram in the way it presents the relevant information. This shows that equations are derivatives or interpretants of diagrams. But diagrams do much more than simply relay information in visual ways. As we shall discuss in this chapter, they are often models of structure that lead to further insights.

The importance of diagrams in semiotic theory has been discussed in detail and need not be revisited here (Saint-Martin 1990; Stjernfelt 2007). Stjernfelt has even labeled this branch of semiotics "diagrammatology". The purpose of this chapter is to look at the role of diagrams as fundamental modeling devices in mathematics. Because a typology of such devices in mathematics would require a volume in its own right, we will limit the discussion here to three basic types of mathematical diagrams – the schema, the graph, and the layout. As Whiteley (2012: 281) has recently written, the study of mathematical modeling brings together several theoretical frameworks, including semiotics and blending theory:

> Mathematical modeling can be viewed as a careful, and rich, double (or multiple) blend of two (or more) significant spaces. In general, modeling involves at least one space that is tangible – accessible to the senses, coming with some associated meaning (semiotics) – and a question to be answered! The mental space also includes a number of features and properties that will, if projected by distracting. The mental space also includes a number of features and properties that, if projected, will support reasoning in the blend. Worse, these "irrelevant" features may suggest alternative blends that are not generative of solutions to the problem. Selective "forgetting" has been recognized as a crucial skill in modeling with mathematics – sometimes referred to as a form of abstraction. Learning which features to project, and which to ignore, is a critical skill [...]

3.2. Diagrams

Simply put, diagrams are visual forms that allow us to model something in its basic form, shape, or structure, without providing realistic detail as in a drawing or photograph. A drawing of triangle, for example, is a *de facto* diagram, showing how the parts hold together:

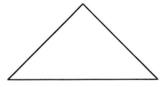

One of the more interesting psychological questions with respect to diagrams is how they are connected to mental images. The research that shows how mental imagery can be elicited is suggestive of this. Among those who have shown a correspondence between our diagrams, maps, and the like and mental imagery is Stephen Kosslyn (1983, 1994), conducting a series of ingenious experiments that have shown how subjects can easily form images in their mind to help them carry out tasks, such as arranging furniture in a room, designing a blueprint, and so on. When people draw maps of their rooms or when architects design building plans there appears to be a high level of consistency between the reported images and the style and form of the diagrams constructed. As Peirce often pointed out, diagrams not only are representamina of external forms, they also mirror the thought processes involved in imagining those forms. In other words, diagrams are themselves models of thought.

Saussure (1916) was among the first human scientists to use the word *image* to refer to what crops up in the brain when a sign is used. However, he never really explored the actual psychological qualities of images. It was with the advent of *image schema* theory, put forward by Lakoff and Johnson (Lakoff, Johnson 1980, 1999; Lakoff 1987; Johnson 1987), that the mental images associated with metaphorical thinking could be compared to the actual diagrams of scientists and others. They define image schemata as largely unconscious mental outlines of recurrent shapes, actions, dimensions, etc. that derive from perception and sensation. Image schemata can, however, be elicited, showing their actual presence in the brain. If someone were to ask for an explanation of the expression "I'm feeling *up* today", one would not likely have a conscious image schema involving an upward orientation. However, if that same person were to be asked the following questions – "How far *up* do you feel?", "What do you mean by *up*?", etc. – then the person would no doubt start to visualize the schema in question.

Image schema theory is thus quite useful in explaining the actual form and style of diagrams, many of which can be defined simply as image schemata that have been given concrete representational form. In a fundamental way, each mathematical diagram is an interpretant (again in the Peircean sense) of a schema. Even the way a problem is presented often betrays the deployment of imagistic thinking. The opposite also seems to be true – that is, an appropriate diagram can often trigger the appropriate schema in the brain that leads to an insight. Whiteley (2012) presents an interesting case in point regarding this. The problem he discusses is:

> *What shape of rectangular box has the maximum volume for a fixed surface area?*

This problem can be easily recast in diagram form as follows: Given a square sheet of material, equal squares are cut out of the corners, and the sides are folded up to make an open box. Which size of corners should be cut out in order to make the maximum volume? Why? The diagram displaying this recasting of the problem is as follows:

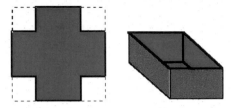

The key step, Whiteley (2012: 264) writes "is to develop a strong sense of what non-optimal boxes look like", which leads to the conclusion that "if the surface area of the sides ≠ the surface area of the base then we do not have the maximum volume". The final critical step is, therefore, that "the surface area of the sides = the surface area of the base". From this, we can see that for four sides to equal the area of the base, each of them should be of the width of the base – or 1/6 the width of the original material.

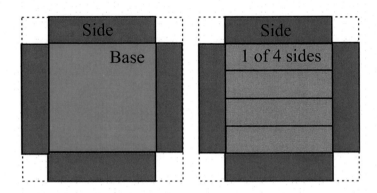

"The solution," concludes Whiteley (2012: 257), "is now clear, sensible – and exact. We have only one location where we do not see how to increase the volume. The blend became rich enough to reason with and confirm, with confidence, the answer which was initially unexpected."

As this simple example shows, diagrams are not just outline representations of raw information. By organizing the information in a schematic way, they reveal the hidden structure in it and thus lead to solutions and new ideas about concepts. They are what Peirce called Existential Graphs, which portray something in its essential form, thus revealing the thought processes used in grasping the form itself.

In every branch of mathematics, diagrams of all kinds are used for a host of reasons – to display data, to represent numbers, to display statistical information, to carry out proofs of theorems, and so on and so forth. Actually, it can be argued that the origin of mathematics as a theoretical discipline with the Pythagoreans lies in perceiving a continuity between numerical and geometrical forms, implying that numbers are themselves counterparts of diagrams. Pythagoras called such numbers *figurate*. For example, square integers, such as 1^2 (= 1), 2^2 (= 4), 3^2 (= 9), and 4^2 (= 16) can be displayed with square arrangements of objects, such as dots:

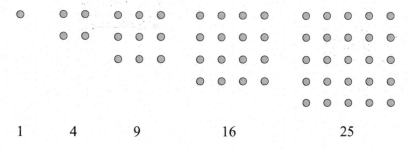

| 1 | 4 | 9 | 16 | 25 |

As a consequence of displaying square numbers in this way, the Pythagoreans discovered that square numbers are equal to the sum of consecutive odd integers:

1	=	1
4	=	1 + 3
9	=	1 + 3 + 5
16	=	1 + 3 + 5 + 7
25	=	1 + 3 + 5 + 7 + 9

And so on.

Why is this so? It is so because to form each new square figure a successive odd number of dots to the preceding figure must be added – for example, to the square figure for the number 1 above 3 new dots must added to produce the next figure for 4; then to the square for the number 4 above, 5 more dots must be added to produce the next figure for 9; and so on. If we examine the square of 25 closely, we can partition it off as follows, using this discovered property:

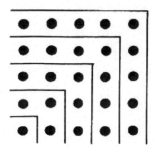

This derived diagram of the same information now shows clearly that 25 is the sum of the first five odd numbers, since it is made up successively of $1 + 3 + 5 + 7 + 9$ additions of dots. By the way, the Pythagoreans called the odd numbers *gnomic* because the partitioning diagram looks like a carpenter's square known as a *gnomon*, which was an L-shaped measuring device. They were intrigued by such findings and, consequently, felt impelled to study all kinds of figurate numbers. The triangular ones held a special kind of place in their scheme of things. Here are the diagrams of the first four triangular numbers:

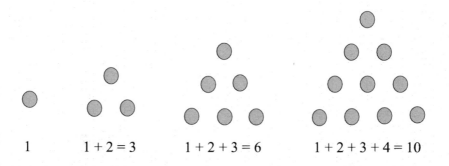

1   1 + 2 = 3   1 + 2 + 3 = 6   1 + 2 + 3 + 4 = 10

As can be seen, the number of dots in the figure for the first triangular figure is 1; the number of dots that makes up the figure for the second triangular number is 3, which is $1 + 2$; the number of dots constituting the third triangular number is 6, which $1 + 2 + 3$; and so on. Thus, each successive triangular number above is obtained by adding a row of dots containing one more than the number of dots in the bottom row of the previous number. This leads to the theorem that the nth triangular number is the sum of the first n counting numbers:

$$
\begin{aligned}
1 &= 1 \\
3 &= 1 + 2 \\
6 &= 1 + 2 + 3 \\
10 &= 1 + 2 + 3 + 4 \\
15 &= 1 + 2 + 3 + 4 + 5 \\
21 &= 1 + 2 + 3 + 4 + 5 + 6 \\
&\ldots \\
n^{th} &= 1 + 2 + 3 + 4 + 5 + 6 + \ldots + n
\end{aligned}
$$

Given their penchant for associating numbers with geometric diagrams, the Pythagoreans also came up with the notions of *amicable* and *perfect* numbers. The numbers 284 and 220, for instance, are called amicable because the proper divisors of one of them when added together produce the other. The proper divisors of 284 are 1, 2, 4, 71, and 142, and the proper divisors of 220 are 1, 2, 4, 5, 10, 11, 20, 22, 44, 55, and 110. Now, we notice that:

$$220 = 1 + 2 + 4 + 71 + 142$$
$$284 = 1 + 2 + 4 + 5 + 10 + 11 + 20 + 22 + 44 + 55 + 110$$

A perfect number, on the other hand, is one that equals the sum of its proper divisors, with the exception of itself. For example, the proper divisors of the number 6 are 1, 2, and 3. Now, if we add these together we get the number 6 = 1 + 2 + 3. The next perfect number is 28. Its proper divisors are 1, 2, 4, 7, and 14, and 1 + 2 + 4 + 7 + 14 = 28. Very few perfect numbers have been discovered. Numbers that are not perfect are called *excessive* or *defective*. An excessive number is one whose proper divisors, when added together, produce a result that exceeds its value. The number 12, for example, is excessive because the sum of its proper divisors, 1, 2, 3, 4, and 6 (1 + 2 + 3 + 4 + 6 = 16) exceeds its value. A defective number is one whose proper divisors, when added together, produce a result that is smaller than its value. One example is 8, since the sum of its proper divisors 1, 2, and 4 (1 + 2 + 4 = 7) is less than its value. Although diagrams were not employed to discuss such numbers, it is our view that the Pythagoreans, who thought diagrammatically, would have come up with these notions without diagrams. The use of diagrams for carrying out many proofs was then adopted by Euclid in his *Elements*. One could even argue that many of the theorems were guided by the structure and form of the diagrams used. Since Euclid, different proofs of the theorems have been used, showing that diagramming is not a fixed system of description; it is an art that allows mathematicians to use their imaginations to draw up models of mathematical information which, in turn, allow them to see things that they would otherwise not have seen.

It is worth revisiting a few simple examples here to illustrate how Euclidean proof and diagrammatic reasoning overlapped. Take, as one example, a Euclidean theorem that involves vertically-opposite angles. These are angles whose vertices touch, as a result of straight lines intersecting. We are required to prove that the angles formed when two straight lines intersect are equal. A simple illustrative diagram reveals the structure of this problem readily:

To prove something as true, Euclid claimed, one must reason in a logical fashion, applying previous knowledge to the problem at hand. To do this, though, requires modifying the representational diagram. One way to do this is by simply numbering the four vertically-opposite angles formed by their intersection as shown. There are, of course, other ways to give appropriate detail to the diagram:

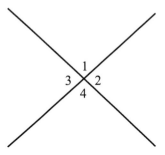

The proof hinges on the established fact that a straight line is an angle of $180°$. We start by considering the line containing 1 + 2. As a straight line, it is an angle of $180°$:

(1) $1 + 2 = 180°$

Now, 1 + 3 also lie on a line and thus also add up to $180°$:

(2) $1 + 3 = 180°$

These two equations can be rewritten as follows:

(1) $2 = 180° - 1$
(2) $3 = 180° - 1$

Since things equal to the same thing are equal to each other we can deduce that 2 = 3. We can now conclude that any two vertically-opposite angles produced by the intersection of two straight lines are equal, because we did not assign a specific value to either angle. We have, in effect, proved a theorem for the general case. The initial diagram and the unfolding of the proof are clearly intrinsically intertwined. One implies the other.

Another classic Euclidean example of diagrammatic reasoning is the formula for the number of degrees in the angles of a polygon. In this proof, it is the triangle form that leads to the formula. We start with the simplest possible case, a triangle itself, which is a polygon with the least number of sides – three. The sum of the angles in a triangle is $180°$:

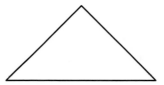

We now go on to consider a quadrilateral (the next case, since it is a polygon with four sides). We start by drawing one, with a diagonal through it. This creates two triangles, as can be seen:

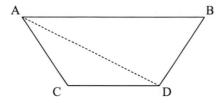

Two triangles can also be produced by joining BC instead. By doing this we have discovered that the sum of the angles in the quadrilateral is equivalent to the sum of the angles in two triangles, namely:

$$180° + 180° = 360°$$

Next, we consider the case of a pentagon (a five-sided figure). ABCDE below is one such figure:

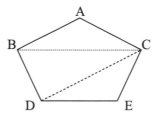

The pentagon can be divided into three triangles, with two diagonals, as shown above (triangle ABC, triangle BCD, and triangle CDE). There are other ways to partition the pentagon, but the result will always be the same – three triangles. We have now discovered another simple fact – the sum of the angles in a pentagon is equivalent to the sum of the angles in three triangles:

$$180° + 180° + 180° = 540°$$

Continuing in this way, it can be shown that the number of angles in a hexagon is equal to the sum of the angles in four triangles (with three diagonals), in a heptagon to the sum of the angles in five triangles (four diagonals), and so on. Let's now generalize what we have apparently discovered. It would appear that the number of triangles that can be drawn in any polygon is two less than the number of sides that define the polygon. For example, in a quadrilateral we can draw two triangles, which is two less than the number of its sides (4), or $(4 - 2)$; in a pentagon, we can draw three triangles, which is, again, two less than the number of its sides (5), or $(5 - 2)$; and so on. In the case of a triangle, this rule also applies, since we can draw in it one and only one triangle (itself). This also is two less than the number of its sides (3), or $(3 - 2)$. In an n-gon, therefore, we can draw $(n - 2)$ triangles:

Number of sides in the polygon	Number of triangles
3 (= triangle)	$(3 - 2) = 1$ triangle
4 (= quadrilateral)	$(4 - 2) = 2$ triangles
5 (= pentagon)	$(5 - 2) = 3$ triangles
6 (= hexagon)	$(6 - 2) = 4$ triangles
7 (= heptagon)	$(7 - 2) = 5$ triangles
...	...
n (= n-gon)	$(n - 2)$ triangles

Since we know that there are $180°$ in a triangle, then there will be $2 \times 180°$ in a quadrilateral, $3 \times 180°$ in a pentagon, and so on. In an n-gon, therefore, there will be $(n - 2) 180°$:

Sides	Triangles	Sum of Degrees
3	$(3 - 2) = 1$	$180° \times 1 = 180°$
4	$(4 - 2) = 2$	$180° \times 2 = 360°$
5	$(5 - 2) = 3$	$180° \times 3 = 540°$
6	$(6 - 2) = 4$	$180° \times 4 = 720°$
7	$(7 - 2) = 5$	$180° \times 5 = 900°$
...	...	...
n	$(n - 2)$	$180° \times (n - 2) = (n - 2) 180°$

Now, with this formula in hand, we can determine the number of degrees in any polygon. Again, it was a basic form of diagrammatic reasoning that led to the discovery of the formula.

3.3. Schemata

According to Lakoff and Johnson, image schemata guide the formation of common concepts. As an example, consider the image schema of an *impediment* (Danesi 2004b). An impediment is something, such as a wall, a boulder, another person, etc. that blocks forward movement. Familiarity with impediments in real life produces a mental outline that can be shown as follows:

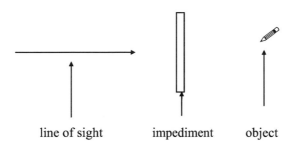

line of sight   impediment   object

As we also know from real experience, we can go *around* an impediment, *over* it, *under* it, *through* it, or else *remove* it and continue on towards the object. On the other hand, the impediment could successfully impede us, so that we would have to *stop* at the impediment and *turn back*. All of these imagined actions can be used to understand the nature of a host of abstract ideas. This is why we say such things as:

> She *got through* that difficult time.
> He felt better after he *got over* his cold.
> You might want to *steer clear of* financial debt.
> With the bulk of the work *out of the way*, he was able to call it a day.
> The rain *stopped* us from enjoying our picnic,
> You cannot *go* any *further* with that idea; you'll just have to *turn back*.

The connection between image schemata and verbal discourse is clearly evident. The schema seems to guide the choice of metaphors in conversations and speech generally. In fact, the pictorial representation above of the impediment schema is a *de facto* diagram of the thought processes involved that produced the speech samples. Without such diagrams Lakoff and Johnson themselves could hardly make an argument for their theory. We can call this type of diagram simply a *schema*. Our use of this term must be clarified. It was perhaps first used by Aristotle and later by Kant in reference to an essential type of reasoning that lays out the argument in a diagrammatic way. For this reason the same term is used in formal logic to refer to the layout of symbols to represent

an argument. The concept of schema used here, however, is consistent with Lakoff and Johnson's theory; we will refer to the Aristotelian schema as a *layout*
below. A schema here is conceived as a diagrammatic counterpart of a mental
image schema.

The use of schematic diagrams is so common and all-encompassing that
we hardly ever realize what it involves. Such diagrams are typically constructed with *points, lines,* and *shapes*. These are the geometric signifiers that can be
combined in various ways to represent specific forms. Consider what can be
done with three straight lines. Among other representations, they can be joined
up to represent a triangle, the letter "H," or a picnic table:

Lines and arrowheads can also be used indexically to represent movement, direction, or orientation:

Shapes are schematic forms designed to show an outline of something. Virtually everything we see can be represented by a combination of lines and shapes:
for example, a cloud can be shown as a shape, a horizon as a line. Other signifiers that go into the make-up of schematic diagrams include *value, color,* and
texture. Value refers to the darkness or lightness of a line or shape. It plays an
important role in portraying dark and light contrasts. Color conveys mood,
feeling, and atmosphere, and in mathematical and scientific diagrams it may
show emphasis or highlight some aspect of the referent. Texture refers to the
sensation of touch evoked synesthetically when we look at some surface. Texture rarely plays a role in mathematical diagrams outside of topological models that often utilize texture to bring out the hidden forms in certain objects.
Lines and shapes can also be combined to create an illusion of depth, as the
art of perspective painting has made saliently obvious. In the following figure
there are 12 lines. The way they are put together, however, makes us believe
that they represent a three-dimensional box or cube:

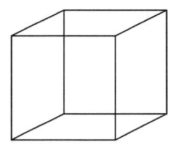

In fact, this perspectival juxtaposition of lines and shapes allows us to draw all kinds of three-dimensional objects on a two-dimensional surface. We can, for instance, easily draw the five Platonic solids (the tetrahedron, the cube, the octahedron, the dodecahedron, and the icosahedron) and view them as if they were three-dimensional:

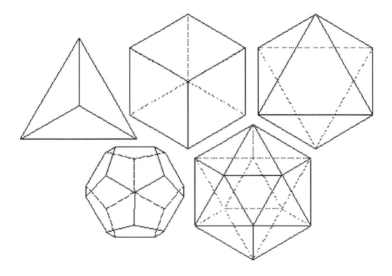

Schematic diagrams do much more than simply reproduce the shape and form of things. They allow us to manipulate shapes and forms intellectually. This is what Euclid and the other Greek geometers did when they added elements to diagrams in order to prove something. As such, they also allow for the exploration of hidden structure. The diagram of the atom as a miniature solar system with a nucleus and orbiting particles is, *ipso facto*, a schematic diagram of atomic structure, allowing us to envision it in a particular way. Atomic structure is unseeable; the diagram gives it a form. It is, of course, an interpretant, not a demonstrable referent. But as such it leads to other visualizations, other models and their diagrams, and so on and so forth. Atomic theory is a diagrammatic theory. The specific type of schema used in science is, thus, often a matter of negotiation and manipulation among scientists.

The concept of schema can be extended to include such forms as equations, since these portray information in a simulative imagistic way. Take, for example, a simple linear equation such as the following one:

$$x + 6 = 2x$$

In its very form, it suggests a balance scale in outline form, with the two sides of the equation corresponding to the two pans on a scale. The equation holds because the value of the terms on the left side must equal the value of the terms on the right side to maintain the balance. This diagram thus translates in visual form the implied image schema of the balance scale. The procedures used to solve equations now can be related to this schema, allowing us to formalize equation-solving. We must subtract the x on the left side, which is the equivalent of removing a weight (x) from the left pan. This means that we must subtract it (remove it) as well from the other side, to keep the balance intact:

$$\overline{(x - x)} + 6 = 2x \overline{- x}$$

Now, $(x - x)$ on the left side gives, of course, the result of 0. One effect of subtracting x from both sides has thus been to eliminate the x from the left side of the original equation:

$$6 = 2x - x$$

Simplifying the expression on the right side gives the result of 1x or simply x. This allows us to rewrite the equation as:

$$6 = x$$
$$x = 6 \qquad \leftarrow \textit{ Turning the sides around}$$

This is, in effect, the solution. The process used to solve it initially is intrinsically diagrammatic since it involves manipulating the image schema of balance scale or some other isomorphic image schema. Interestingly, relevant studies have found that children can solve equations when they are cast in schematic-diagrammatic form, because these are perceived to have real-life meaning (Liebeck 1984; Radford, Grenier 1996; Swafford, Langrall 2000; Musser et al. 2006). Diagrammatically, it can also be seen that the end result of subtracting x from both sides was to "move the x on the left side over to the right side" changing its sign in the process:

$$x + 6 = 2x$$
$$6 = 2x - x$$

This procedure allows us to skip the step that eliminates a term on either side. It suggests a general procedure for *transposing* terms from side to side, as it is called: that is, we can eliminate something from any side of the equation by "bringing it over" to the other side, changing its sign in the process. This algorithm is, clearly, the result of a previous diagrammatic thought process that involved the image schema of a scale. It can be claimed that mathematics works this way at all levels and in all domains. First, some schematic model is used to set up something and then the schema suggests how it is to be carried out or what it implies. Once this is discovered it can be systematized algorithmically, with the original schema becoming unconscious. At different levels, actually, the original form may have further diagrammatic manifestations. For example, as Cartesian geometry has shown, equations encode geometric information. An equation such as the one above is called *linear* because, if plotted on graph paper, the equation produces a line. This also leads to a further insight – namely, that equations of the first degree (with the variable to the first power) are all linear, because they produce linear schemata. Equations of different degrees produce different geometric forms. In fact, Cartesian geometry has simply given the Pythagorean principle of numbers as diagrams a different twist, showing that when put into equations, numbers produce geometric forms.

As one more example of the role of schematic diagrams in the development of mathematical ideas, consider the emergence of graph theory – a branch that is traced to a simple problem, namely Euler's Königsberg bridges problem, which he formulated in a famous 1736 paper titled "*Solutio problematis ad geometriam situs pertinentis*" ("The solution of a problem relating to the geometry of position"). He presented his solution first to the Academy in St. Petersburg, Russia. He published it later in 1741.

In the old Prussian town of Königsberg runs the Pregel River, in which there are two islands. In Euler's times they were connected with the mainland and with each other by seven bridges. The residents of the town would often debate whether or not it was possible to take a walk from any point in the town, crossing each bridge once and only once, and returning to the starting point. No one had found a way to do it but, on the other hand, no one could explain why it seemed to be impossible. Euler became intrigued by the debate, turning it into one of the greatest problems of all time, since it produced not only a new branch of mathematics, but also many new insights into the relation between numbers and visual models.

A schema diagram, rather than a reproduction of the scene, will bring out the structural features of the situation, since it presents the information in outline form, highlighting its critical structural features apart from the clutter of other features:

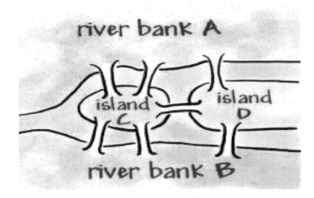

This map can now be further schematized just highlighting the land regions represented with capital letters (A, B, C, D) and the bridges with lower-case letters (a, b, c, d, e, f, g):

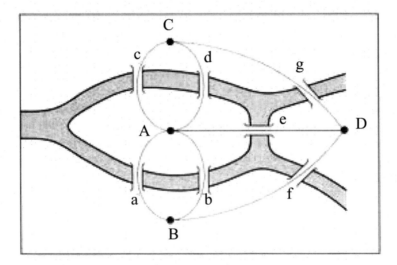

This new schematic version of the problem provides an even more manageable depiction of the situation because it disregards the distracting shapes of the land masses and bridges, reducing them to *points* or *vertices*, and portraying the bridges as *paths* or *edges*. This is called a *network* in contemporary graph theory. The above diagram can be rendered even more schematic by reducing its structural properties into a simple model:

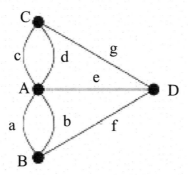

This diagram now represents the original problem as a simple network schema that shows the solution visually. It shows in itself that it is impossible to cross over the bridges just once without doubling back. The relevant point to be made here is that from this problem and its diagrammatic solution, two whole new branches of mathematics emerged – graph theory and later on topology. It is relevant to note that Euler himself apparently never used a diagram in his initial paper. He just described it in words, which is an amazing feat in itself, betraying a huge mathematical imagination. Richeson (2008: 104) clarifies the situation as follows:

> It is a common misconception that the Königsberg graph is found in Euler's paper. In reality, neither the Königsberg graph nor any graph appears there. Graph tracing developed independently of the Königsberg bridge problem. Graph tracing puzzles appeared in the early nineteenth century, both in mathematical articles and in books of recreational mathematics. It was not until 1892 that W. W. Rouse Ball (1850–1925) in his popular work *Mathematical Recreations and Problems*, made the connection between Euler's result on the bridges of Königsberg and graph tracing. The first appearance of the Königsberg graph was in Ball's book, over a hundred and fifty years after the publication of Euler's paper.

Nonetheless, a drawing of the problem has been found among Euler's papers (Flood, Wilson 2011: 117), which undoubtedly guided Euler's solution. What this episode in the history of mathematics shows is that diagrammatic reasoning is crucial to conducting explorations within the discipline and that it often leads to discoveries.

3.4. Graphs

In mathematical circles, Euler's diagram is called a *graph* rather than a *schema*. Actually, the term graph is an appropriate one, constituting a more general term. It is a kind of schema that brings out structural features of a situation graphically, and is now limited in meaning within the field to denote a diagram showing vertices, paths, and so on. This is why Euler's discovery led to the branch called *graph theory*, which looks at the relationships among these elemental geometric forms:

> The solution to the Königsberg bridge problem illustrates a general mathematical phenomenon. When examining a problem, we may be overwhelmed by extraneous information. A good-problem-solving technique strips away irrelevant information and focuses on the essence of the situation. In this case details such as the exact positions of the bridges and land masses, the width of the river, and the shape of the island were extraneous. Euler turned the problem into one that is simple to state in graph theory terms. Such is the sign of genius. (Richeson 2008: 107)

Euler himself discovered several fundamental properties about graphs. In the case of a three-dimensional figure, for instance, he found that if we subtract the number of edges (e) from the number of vertices (v) and then add the number of faces (f) we will always get 2 as a result:

$$v - e + f = 2$$

Take, for example, a cube. It has 8 vertices, 12 edges, and 6 faces. Now, inserting these values in the formula, it can be seen that the relation it stipulates holds:

$$v - e + f = 2$$
$$8 - 12 + 6 = 2$$

Euler also proved that for plane figures the value of $v - e + f$ is 1 rather than 2. A rectangle, for instance, has 4 vertices, 4 edges and 1 face. Therefore:

$$v - e + f = 1$$
$$4 - 4 + 1 = 1$$

The Königsberg Bridges graph, being a planar graph, also possesses this property. It has 4 vertices, 7 edges, and 4 faces. Therefore:

$$v - e + f = 1$$
$$4 - 7 + 4 = 1$$

From these simple observations, graph theory has led to so many discoveries and ideas in mathematics that it would take a huge tome on its own just to list them. As Richeson (2008: 9) observes, a simple observation such as that by Euler has led to so many discoveries and branches of mathematics that it boggles the mind:

> Euler's formula should be as well known as these great theorems [the infinitude of primes, the irrationality of π]. It has a colorful history, and many of the world's greatest mathematicians contributed to the theory. It is a deep theorem, and one's appreciation for this depth grows with one's mathematical sophistication.

The concept of graph was a primary one in Peirce as well, who put forward a system of diagrammatic logic that he called Existential Graphs (CP 4.347–584). Like Euler, Peirce saw a graph as anything having its parts in relation to each in such a way that they resemble relations among the parts of some different set of entities or referents. A graph is, thus, a schematization of these relations. It is a pictorial manifestation of what goes on in the mind imagistically. Graphs thus display the very process of thinking *in actu* (CP 4. 6), showing how a given mathematical argument unfolds in a visual way. They thus allow us to grasp something as a set of transitional states, where the meaning of the referential domain is accessible in its entirety in the visual form of the graph itself. Therefore, every graph conveys information and simultaneously explains how it is being done. It is a picture of cognitive processes in action. The following citation encapsulates Peirce's notion of graph. In it, we see him discussing with a general why a map is used to conduct a campaign:

> But why do that [use maps] when the thought itself is present to us? Such, substantially, has been the interrogative objection raised by an eminent and glorious General. Recluse that I am, I was not ready with the counter-question, which should have run, "General, you make use of maps during a campaign, I believe. But why should you do so, when the country they represent is right there?" Thereupon, had he replied that he found details in the maps that were so far from being "right there," that they were within the enemy's lines, I ought to have pressed the question, "Am I right, then, in understanding that, if you were thoroughly and perfectly familiar with the country, no map of it would then be of the smallest use to you in laying out your detailed plans?" No I do not say that, since I might probably desire the maps to stick pins into, so as to mark each anticipated day's change in the situations of the two armies." "Well, General, that precisely corresponds to the advantages of a diagram of the course of a discussion. Namely, if I may try to state the matter after you, one can make exact experiments upon uniform diagrams; and when one does so, one must keep a bright lookout for unintended and unexpected changes thereby brought about in the relations of different significant parts of the diagram to one another. Such operations upon diagrams, whether external or imaginary, take the place of the experiments upon real things that one performs in chemical and physical research." (CP 4.530)

Peirce makes it clear that graphs and cognition (including language) are mirrors of each other. But maps also allow us to experiment with the object of reference, thus allowing us to infer things that may not be evident in other forms. To put it differently, graphs contain information that in itself leads to further insight. In fact, several decades after Euler's proof of the Königsberg problem, mathematicians began studying figures that retained their structural features after being deformed. Their efforts led to the establishment of yet another branch of mathematics called *topology*. The first comprehensive treatment of the field, titled *Theory of Elementary Relationships*, was published in 1863. It was written by the German mathematician Augustus Möbius (1790–1868), the inventor of a truly enigmatic figure called the Möbius Strip.

Topology concerns itself with determining such things as the *insideness* or *outsideness* of shapes. A circle, for instance, divides a flat plane into two regions, an inside and an outside. A point outside the circle cannot be connected to a point inside it by a continuous path in the plane without crossing the circle's circumference. If the plane is deformed, it may no longer be flat or smooth, and the circle may become a crinkly curve, but it will continue to divide the surface into an inside and an outside. That is its defining structural feature. Topologists study all kinds of figures in this way. They investigate, for example, knots that can be twisted, stretched, or otherwise deformed, but not torn. Two knots are equivalent if one can be deformed into the other; otherwise, they are distinct. Topology also came about as mathematicians studied Eulerian and non-Eulerian forms (that is those that had the Eulerian structure as expressed by his formulas, and those that did not). Topology then generalized Eulerian graph theory. Richeson (2008: 155) puts it as follows:

> The fruitful dialogue about Eulerian and non-Eulerian polyhedra in the first half of the nineteenth century set the stage for the field that would become topology. These ideas were explored further by others, culminating in Poincaré's marvelous generalization of Euler's formula at the end of the nineteenth century.

Topology makes no distinction between a sphere and a cube because these figures can be molded into one another. It does make a distinction between a sphere and a *torus* (a doughnut-shaped figure) because a sphere cannot be deformed into a torus without being torn. Interestingly, from topological theory has come an interest in knots as models of natural phenomena, showing again the principle enunciated in this book that artifactual models often contain within them hidden factual models of the world, revealing more about physical reality than we could have ever imagined. It appears, for example, that the DNA and the universe itself may have topological structure of a certain kind. Stewart (2012: 105) elaborates as follows:

One of the most fascinating applications of topology is its growing use in biology, helping us understand the workings of the molecule of life, DNA. Topology turns up because DNA is a double helix, like two spiral staircases winding around each other. The two strands are intricately intertwined, and important biological processes, in particular the way a cell copies its DNA when it divides, have to take account of this complex topology.

This interaction between artifactual models and natural ones will be taken up again in the final chapter. Suffice it to say here that it seems to be a semiotic fact of life, which Lotman often articulated as the semiosphere overlapping with the biosphere. Thus even highly abstract models of mathematical structure, with no apparent applications, turn out to be applicable to reality eventually. To quote Richeson (2008: 201):

> But even the most abstract and theoretical areas of mathematics can prove useful. Applicable mathematics often comes from decidedly non-applicable areas. The usefulness of a particular theory is often not clear for many years. No one could have predicted that the study of prime numbers would later enable us to encrypt credit-card information so that it can be sent safely across the Internet. Nineteenth-century mathematicians did not know that their work in non-Euclidean geometry would provide the foundation for Einstein's theory of general relativity. Toward the end of the nineteenth century the usefulness of knot theory reemerged in the natural sciences. Physicists, biologists, and chemists discovered that the mathematical theory of knots gave them insight into their fields. Whether it is the study of DNA or other large molecules, magnetic field lines, quantum field theory, or statistical mechanics, knot theory now plays an important role.

To summarize, the mathematician's diagrams often become the scientist's theories. As the great French mathematician Poincaré pointed out, he came to his own discoveries by considering the simple diagrams of the two Newtonian curve models of orbits: "When one tries to depict the figure formed by these two curves and their infinity of intersections, each of which corresponds to a doubly asymptotic solution, these intersections form a kind of net, web or infinitely tight mesh. One is struck by the complexity of this figure that I am not attempting to draw" (cited in Stewart 2012: 64). As this implies, scientists use diagrams to model physical phenomena. But these would be simplistic pictures unless they are tied to mathematical diagrammatic form reasoning. When they are, they allow the scientist to translate a natural phenomenon into equations and other modeling devices. As Banks (1999: 183) puts it:

> Over the centuries, just about everything involving engineering, science, and technology – from Egyptian pyramids and medieval cathedrals to supersonic aircraft and maglev trains – has utilized mathematics to one extent or another.

3.5. Layouts

Recall Eratosthenes's sieve diagram in the previous chapter, which was a layout of the first 100 numbers in the form of an array so that the numbers that are not crossed out can be thought of as having passed through a sieve (strainer) that has caught all the primes. The sieve ends up "trapping" 25 primes. Needless to say, it would be an unwieldy task to set up a sieve to identify the primes even among the first 1000 numbers. But the sieve diagram suggests a more generic problem: can some kind of diagram or model be drawn to help us detect higher primes? As it turns out, this is still one of the great unsolved problems of mathematics that can be reformulated in different terms: is there a formula or algorithm for generating primes, and only primes?

Eratosthenes's diagram can be called simply a layout. Layouts have been instrumental in the discovery of many mathematical ideas and principles. Two of the most famous ones, already discussed in this book, are the layout of the numbers by Cantor and his putting them into a one-to-one relation with other numerical entities and Pascal's triangle, which resulted from a layout of the binomial expansion. Both layout strategies led to considerable discoveries in mathematics, as we saw.

Another classic example of the power of the layout to reveal structure comes from a tale that is told about the German mathematician Karl Friedrich Gauss (1777–1855), who was only nine years old when he purportedly dazzled his teacher, a certain J. G. Büttner, with his exceptional mathematical abilities. One day, his class was asked to cast the sum of all the numbers from one to one hundred: $1 + 2 + 3 + 4 + ... + 100 = ?$ Gauss raised his hand within seconds, giving the correct response of 5,050, astounding both his teacher and the other students who continued to toil over the seemingly gargantuan arithmetical task before them. When his teacher asked Gauss how he was able to come up with the answer so quickly, he is said to have replied (more or less) as follows:

> There are 49 pairs of numbers that add up to one hundred: $1 + 99 = 100$, $2 + 98 = 100$, $3 + 97 = 100$, and so on. That makes 4,900, of course. The number 50, being in the middle, stands alone, as does 100, being at the end. Adding 50 and 100 to 4,900 gives 5,050.

Impressed, the teacher not only arranged for Gauss's admittance to a school with a challenging curriculum, but he also secured a tutor and advanced textbooks for the brilliant child. Let's take a close look at what Gauss saw in his mind. First, he laid out the numbers into "half sequences": (1) from 1 to 49, eliminating 50 from this half, and (2) the remaining numbers from 51 to 99, eliminating the last number 100.

First half	Mid Point	Second half
↓	↓	↓
1, 2, 3, 4, 5, ... 49	50	51, 52, ... 95, 96, 97, 98, 99

He then added together the first number in the first half and last number in the second half, the second in the first half and the second-last in the second half, the third in the first half and third-last in the second half, and so on. He did this because he saw that this pattern produced the constant sum of 100:

1 + 99	=	100	←	*First and last numbers*
2 + 98	=	100	←	*Second and second-last numbers*
3 + 97	=	100	←	*Third and third-last numbers*
4 + 96	=	100	←	*Fourth and fourth-last numbers*
5 + 95	=	100	←	*Fifth and fifth-last numbers*

...

Now, how many such pairs are there in total? As Gauss pointed out to his teacher, the pairings will end at 49 (which is paired with 51). This means that there are 49 pairs that add up to 100. The sum of these pairs is thus 49 × 100 = 4,900. Adding to this the 50 and 100 that he had taken out of the sequence makes 4,900 + 50 + 100 = 5050.

Gauss thus used a layout form that revealed a hidden structure in addition. The mark of the true mathematician is the ability to envision short cuts – ways of doing mathematics that render it efficient and, at the same time, interesting. But that's not all the layout does – Gauss's marvelous insight led to the concept of sequences and, subsequently, to a new area of mathematical study. Using the language of algebra, the problem that Gauss solved can be expressed more generally as: what is the sum of $\{1 + 2 + 3 + ... + n\}$, where n is any whole number? The answer is: $n(n + 1)/2$. Substituting 100 for n produces the answer 5,050.

The examples mentioned here – Eratosthenes' sieve, Cantor's numerical layout, Pascal's triangle, Gauss's sequence – just scratch the surface of how the layout diagram has helped mathematicians do things not only more efficiently, but also discover hidden truths. More generally, it can be said that more often than not a diagrammatic model of a problem is the only way to understand or grasp a pattern in it. As mentioned, diagrams are not fixed representations. They vary from mathematician to mathematician. As Whorf pointed out in many of his writings on linguistics, all science is really a subjective art, with diagrams in linguistics or mathematics revealing the *forma mentis* of the diagram-maker. After all, who could have come up with Cantor's diagonal proof other than Cantor? And often the type of diagrammatic logic used will depend on where the mathematician has lived and the experiences he brings from real life. Whorf (2012: 154) puts it as follows with regard to the linguist:

This fact is very significant for modern science, for it means that no individuals are free to describe nature with absolute impartiality but are constrained to certain modes of interpretation even while they think themselves most free. The person most nearly free in such respects would be a linguist familiar with very many widely different linguistic systems. As yet no linguist is in any such position. We are thus introduced to a new principle of relativity, which holds that all observers are not led by the same physical evidence to the same picture of the universe, unless their linguistic backgrounds are similar, or can in some way be calibrated.

Someone else could have come up with infinite set theory other than Cantor, but it would have been different. But Cantor's background, his individual ingenuity, the kinds of historical antecedents he was exposed to, and so on, produced an interpretant that he translated into a layout diagram. From there, he was able to amplify his diagrammatic methods and discover even more startling truths.

In the end, mathematics is nothing more than a human invention, a particular kind of modeling system that we utilize to understand the world, the umwelt, in terms of our particular type of *Innenwelt*. The axioms and postulates of the founders of mathematics as a theoretical enterprise saw the power of visual thinking in bringing about the discovery of knowledge. But the knowledge gained, as Gödel showed, will always be imperfect and mathematicians will have to live with this fact. Crilly (2011: 11–12) puts it as follows:

The Greeks assumed the truth of their axioms, but today's mathematicians expect only that axioms be consistent. In the 1930s Kurt Gödel rocked mathematics when he proved his incompleteness theorem, which held that there were some mathematical statements in a formal axiomatic system that could neither be proved or disproved using only the axioms of the system. In other words, mathematics could now contain unprovable truths that might just have to stay that way.

4. Blends and connective modeling in mathematics

4.1. Introduction

As mentioned briefly in the opening chapter, it was Max Black (1962) who put forward the notion that scientific models are basically metaphors or the result of metaphorical thinking. Science, as Black elaborated, is essentially an attempt to render visible those things we can never see with our eye – atoms, sound waves, gravitational forces, magnetic fields, etc. The trace to this "inner vision" is metaphor. This is why sound waves are said to *undulate* through empty space like water waves ripple through a still pond; atoms to *leap* from one quantum state to another; electrons to *travel in circles* around an atomic nucleus; and so on. The physicist K. C. Cole (1984: 156) puts it as follows:

> The words we use are metaphors; they are models fashioned from familiar ingredients and nurtured with the help of fertile imaginations. "When a physicist says an electron is like a particle", writes physics professor Douglas Giancoli, "he is making a metaphorical comparison like the poet who says "love is like a rose". In both images a concrete object, a rose or a particle, is used to illuminate an abstract idea, love or electron.

As scientist Roger Jones (1982: 4) has also pointed out, for the scientist metaphor serves as "an evocation of the inner connection among things." The philosopher of science Fernand Hallyn (1990) has identified the goal of science as that of giving the world a "poetic structure". In other words, a scientific or mathematical theory is a kind of poetic-metaphorical model. For example, in physics it was at first speculated that atomic structure mirrors the solar system – a theory that reified the ancient Greek concept of the cosmos as having the same structure at all its levels, from the microcosmic (the atom) to the macrocosmic (the universe). It is mind-boggling to think that such a simple conceptual linkage has led to real knowledge about atoms. Physicists have never seen the inside of an atom with their eyes (nor has anyone else for that matter). So, they use their inner metaphorical eye to produce a hunch that the atom has the same kind of structure that the solar system has, with electrons behaving like little planets orbiting around an atomic nucleus. This model of atomic structure as a miniature solar system continues to make sense because we feel that the cosmos is structurally the same at all levels, from the microcosmic (the atomic) to the macrocosmic (the universe). And, of course, as

Black pointed out, this is so because we perceive the atom and the solar system to be connected. This very metaphor became a model in physics that led to a whole spate of experimental research and discoveries that now fall under the rubric of quantum physics. Physicists no longer use the simple metaphor, but without it they never would have come up with the ideas that they have.

All this raises several rather profound philosophical questions, which we will entertain in the next chapter. In this one, we will focus on metaphorical modeling in mathematics and what it implies. As Lakoff and Núñez (2000) have argued in their controversial book, metaphor plays a fundamental role in mathematical cognition and in the discovery of concepts in the field. Through the analysis of basic concepts from simple counting to the infinitesimal calculus, Lakoff and Núñez have put forward the claim that mathematics is brought forth, not through abstract contemplation, but through the recruitment of everyday cognitive mechanisms that make human imagination and abstraction possible, such as metaphors and conceptual blends. Gilles Fauconnier and Mark Turner (2002, 2008) have proposed arguments along the same lines. Data and emerging results in this domain of inquiry have been giving substance to the notion that metaphorical models are, in fact, crucial in mathematics. Actually, in the loose definition of metaphor used now within cognitive science, it can be claimed that every mathematical model is a metaphor of sorts, an abduction, as Peirce called it, based on inferences deriving from experiences and associations within these experiences. In MST, this is really nothing new, since metaphorical models are forms that result from connective reasoning (Sebeok, Danesi 2000), that is from connecting forms (signifiers) to meanings and vice versa and then connecting them textually to produce forms such as diagrams, equations, and the like.

4.2. Metaphor

Before discussing the role of metaphor in mathematical modeling, it is worthwhile here to go over some of the main findings and notions of what has come to be known as Conceptual Metaphor Theory (CMT). This brief excursion is intended to lay the theoretical groundwork for the remainder of the discussion in this chapter.

It is relevant to note that metaphor has traditionally been relegated to the purely poetic domain, starting with Aristotle, the originator of the term, with literal language being considered the backbone of language and cognition in Western philosophy and linguistics. Some have even viewed metaphor as a defect of human reasoning. The source of latter view is, probably, John Locke's (1975[1690]: 34) characterization of metaphor as a "fault" in his *Essay Concerning Human Understanding*:

> If we would speak of things as they are, we must allow that all the art of
> rhetoric, besides order and clearness, all the artificial and figurative applica-

tion of words eloquence hath invented, are for nothing else but to insinuate wrong ideas, move the passions, and thereby mislead the judgment; and so indeed are perfect cheats: and therefore, however laudable or allowable oratory may render them in harangues and popular addresses, they are certainly, in all discourses that pretend to inform or instruct, wholly to be avoided; and where truth and knowledge are concerned, cannot but be thought a great fault, either of language or person that makes use of them.

Thomas Hobbes also inveighed fiercely against metaphor, characterizing it as an obstacle to understanding, a source of ambiguity and obscurity, and thus, a feature of language to be eliminated from true philosophical and scientific discourse. Hobbes (1839[1656–58]) came to possess this view of metaphor because he believed that the laws of arithmetic mirrored the laws of human thought, and thus that the only meaningful form of philosophical discourse was of the same "literal" kind as the one used to explicate mathematical notions, at least in his own understanding of language and mathematics. It is relevant to note that not everyone held this view. In his *Summa Theologica* (1266–1273), St. Thomas argued that the writers of Holy Scripture presented "spiritual truths" under the "likeness of material things" because that was the only way in which humans could grasp such truths, thus implying that metaphor was a tool of cognition, not just a feature of rhetorical flourish:

> It is befitting Holy Scripture to put forward divine and spiritual truths by means of comparisons with material things. For God provides for everything according to the capacity of its nature. Now it is natural to man to attain to intellectual truths through sensible things, because all our knowledge originates from sense. Hence in Holy Scripture spiritual truths are fittingly taught under the likeness of material things. (quoted in Davis, Hersh 1986: 250)

But despite St. Thomas' discerning observation, philosophers continued largely to ignore metaphor as a cognitive force in the construction of scientific and mathematical models. It was the Neapolitan philosopher Giambattista Vico who attempted to spark interest in it over four centuries later, emphasizing that metaphor was indeed evidence of how "knowledge originates from sense", as St. Thomas had so aptly phrased it. Vico's characterization of our sense-making capacity as "poetic logic" *(logica poetica)* or "poetic wisdom" *(sapienza poetica)* is the first modern psychological theory of metaphor. Incredibly, it continues to remain marginal among psychologists and linguists. Discussing the reasons for this are beyond the objectives of the present discussion (see, for instance, Danesi 1993). Vico (1984[1725]: 122) saw metaphor as an imaginative strategy for understanding what is unknown by enlisting what is familiar:

"It is another property of the human mind that whenever men can form no idea of distant and unknown things, they judge them by what is familiar and at hand". This is not just a matter of convenience or expedience. Rather, it is the only way new ideas are formed. A little later, Immanuel Kant (1965[1781]) also mentioned in his *Critique of Pure Reason* that figurative language was evidence of how the mind attempted to understand unfamiliar things, and Friedrich Nietzsche (1979[1873]) saw metaphor as humanity's greatest flaw, because of its subliminal power to persuade people into believing it on its own terms.

Modern-day scientific interest in metaphor starts with the work of the early experimental psychologists in the latter part of the nineteenth century. The founders of the new discipline were the first to conduct experiments on how people processed figurative language (Wundt 1901; Bühler 1934, 1951[1908]; Stählin 1914). Bühler (1905[1908]), for instance, collected some intriguing data on how subjects paraphrased and recalled proverbs. He found that the recall of a given proverb improved if it was linked to a second proverb; otherwise the proverb was easily forgotten. Bühler concluded that metaphorical-associative thinking produced an effective retrieval form of memory and was, therefore, something to be investigated further by the fledgling science.

Shortly after Bühler's fascinating work, the *Gestalt* movement emerged to make the study of metaphor a primary target of research from the 1920s to the mid-1960s (Wertheimer 1923; Asch 1950, 1958; Osgood, Suci 1953; Brown et al. 1957; Werner, Kaplan 1963; Koen 1965). Asch (1950), for instance, examined metaphors of sensation (*hot, cold, heavy*, etc.) in several unrelated languages as descriptors of emotional states. He found that *hot* stood for *rage* in Hebrew, *enthusiasm* in Chinese, *sexual arousal* in Thai, and *energy* in Hausa (a language spoken in northern Nigeria, Niger, and adjacent areas). This suggested to him that, while the specific emotion implicated varied from language to language, the metaphorical process did not. Simply put, people seemed to think of emotions in terms of physical sensations and expressed them as such.

The scholar who kindled a broad scientific interest in metaphor was, in actual fact, not a psychologist, but a literary theorist. In his groundbreaking 1936 book, *The Philosophy of Rhetoric*, I. A. Richards started a veritable revolution in thought by arguing persuasively that metaphor could not be classified simply as a replacement of literal meaning for decorative or stylistic purposes, but rather that it produced a new psychologically powerful form of meaning that could never be encompassed by a simple literal paraphrase. In order to discuss how such meaning arose, Richards labeled the parts of the metaphor as follows: (1) he called the metaphorical concept itself the *tenor*; (2) he called the concrete notion to which it was linked the *vehicle*; and (3) he called the meaning produced by the linkage of the topic and vehicle the *ground*. Richards then claimed that the linkage was hardly a matter of mere comparison or substitution, but rather an entailment based on perception of relationship. By calling it the *ground*,

Richards wanted clearly to imply that the topic and vehicle stood in relation to each other as do the figures in a painting. Black (1962), as mentioned, followed this up with an in-depth study of the role of metaphor in generating new insight and knowledge. In Black we have the first mention of metaphor as a modeling device – a primary strategy of encoding new knowledge on the basis of how it appears to the model-maker's experience. Each theory is a metaphor, an implicit image schema of some phenomenon. This is why every scientific theory is also a diagram or graph in the Peircean sense (Chapter 3).

The final turning point was a 1977 study, *Psychology and the Poetics of Growth: Figurative Language in Psychology, Psychotherapy, and Education*, conducted by a research team of psychologists headed by Howard Pollio, which showed that metaphor pervades common everyday speech (Pollio et al. 1977). The team found that speakers of English uttered, on average, an astounding 3,000 novel metaphors and 7,000 idioms per week. In other words, the study made it obvious to everyone that metaphor could hardly be construed as a deviation from linguistic rules, or a mere stylistic option to literal language. Since then, the number of volumes, symposia, courses, articles, websites, and journals dealing directly with, or involving related work on, metaphor has become astronomical. We cannot possibly give even a schematic overview here of the relevant literature.

The 1979 collection of studies edited by Andrew Ortony, *Metaphor and Thought*, and the 1980 anthology put together by Richard P. Honeck and Robert R. Hoffman, *Cognition and Figurative Language*, along with the 1980 book by George Lakoff and Mark Johnson, *Metaphors We Live By*, set the groundwork for a new approach to the study of language and cognition that has, as mentioned, since come to be known as *Conceptual Metaphor Theory* (CMT). The central notion on which CMT is implanted is that metaphorical meaning pervades language and thought. The traditional literalist approach asserts that we encode and decode linguistic messages on the basis of literal meaning, deciding on a metaphorical one only when a literal interpretation is not possible. But CMT has brought forward an enormous amount of evidence to show why this view is no longer tenable. If contextual information is missing from an utterance such as "The murderer is an animal", our inclination is to interpret it metaphorically, not literally. It is only if we are told that the *murderer* is an actual "animal" (a bear, a cougar, etc.) that a literal interpretation comes into focus.

In their groundbreaking work, Lakoff and Johnson (1980) asserted at first what Aristotle claimed two millennia before, namely, that there are two types of concepts – *concrete* and *abstract*. But the two scholars added a twist to this dichotomy, suggesting that abstract concepts should not be viewed as being autonomous from concrete ones, but rather as being built up systematically from them. They renamed an abstract concept a *conceptual metaphor*. To grasp what this implies, consider the linkage of humans with animals: "John is a gorilla",

"She is a snake", etc. These are specific *linguistic metaphors*, but they are obviously interconnected, revealing, in fact, a more general conceptual metaphor: *people are animals*. The two scholars then labeled each of the two parts of a conceptual metaphor *domains* – they called the *people* part the *target domain* because it is the abstract "target" of the conceptual metaphor; and they termed the *animals* part of the metaphor the *source domain* because it constitutes the class of vehicles that delivers the metaphor (the "source" of the metaphorical concept). The gist of Lakoff and Johnson's argument is that the specific linguistic metaphors people use commonly in everyday conversations are hardly disconnected flights of linguistic fancy, but rather manifestations of interconnected thinking that crystallizes systematically in speech.

CMT has radically altered the ways in which semantic systems are studied today by a large number of psychologists and linguists. Since ancient times metaphor was studied as a rhetorical device, rather than as a trace to human thought. In rhetorical tradition, metaphor is still viewed as one of various tropes. The trend within CMT is to consider most tropes (except for metonymy and irony) as manifestations of metaphorical reasoning, rather than as separate tropes. Thus, for example, personification ("My cat speaks Italian", "Mystery resides here") is now viewed as a specific kind of conceptual metaphor, in which *people* is the source domain. In all cases, a mapping of one domain onto the other defines the process. Lakoff (2012: 157) puts it as follows:

> The word metaphor has come to mean a cross-domain mapping in the conceptual system. The term metaphorical expression refers to a linguistic expression (a word, phrase, or sentence) that is the surface realization of such a cross-domain mapping (this is what the word metaphor referred to in the old theory).

There are a few caveats that must be made from the outset vis-à-vis CMT. First, whether or not abstract concepts are structured metaphorically (or metonymically) is a question that is open to research and debate. Second, even if this were so, it must not be forgotten that there are aspects of language and cognition that are not metaphorical or figurative (indexical and symbolic, for instance). But despite these cautions, the current research on the comprehension and production of metaphor within CMT has made it no longer tenable to assign figurative meaning to some subordinate category vis-à-vis literal meaning. In fact, metaphor is so common in discourse and cultural representation that we hardly ever realize how it influences our perceptions and beliefs.

4.3. Metaphorical modeling

The idea that metaphor plays a role in mathematical logic seems to have never been held seriously until very recently. As Solomon Marcus (2012: 124) observes, this is so despite the fact that mathematical terms are themselves metaphors:

> For a long time, metaphor was considered incompatible with the requirements of rigor and preciseness of mathematics. This happened because it was seen only as a rhetorical device such as "this girl is a flower." However, the largest part of mathematical terminology is the result of some metaphorical processes, using transfers from ordinary language. Mathematical terms such as *function, union, inclusion, border, frontier, distance, bounded, open, closed, imaginary number, rational/irrational number* are only a few examples in this respect. Similar metaphorical processes take place in the artificial component of the mathematical sign system.

It was Lakoff and Núñez's (2000) book that introduced metaphorical modeling to mathematicians. Consider a simple statement such as "7 is larger than 4". This is hardly a literal pronouncement. Rather it reveals a source domain that involves size while the target domain entails numbers, as Presmeg (1997, 2005) also claims. The basic metaphorical concept that underlies such a statement is that *numbers are collections of objects of the same* size. This kind of analysis can be applied throughout mathematics, according to Lakoff and Núñez. The concept of quantity, for instance, involves at least two metaphors. The first is the *more is up, less is down* conceptual metaphor, which appears in common expressions such as *prices went up* and *the stock market plummeted*. The other is that *linear scales are paths,* which manifests itself in expressions such as *rational numbers are far more numerous than integers,* and *infinity is way beyond any collection of finite sets.* As Lakoff (2012: 164) puts it:

> The metaphor maps the starting point of the path onto the bottom of the scale and maps distance traveled onto quantity in general. What is particularly interesting is that the logic of paths maps onto the logic of linear scales. Path inference: If you are going from A to C, and you are now at intermediate point B, then you have been at all points between A and B and not at any points between B and C. Example: If you are going from San Francisco to N.Y. along Route 80, and you are now at Chicago, then you have been to Denver but not to Pittsburgh. Linear scale inference: If you have exactly $50 in your bank account, then you have $40, $30, and so on, but not $60, $70, or any larger amount. The form of these inferences is the same. The path inference is a consequence of the cognitive topology of paths. It will be true of any path image-schema.

Lakoff and others in the CMT movement have proposed (as mentioned) that conceptual metaphors are formed in the brain through a process called *blending*. Thus, a simple statement such as "7 is larger than 4" is really a conceptual blend that takes the concept of *numbers* in one domain and maps it with *objects* from another, producing the conceptual metaphor *numbers are collections of objects* (that is, numbers in themselves are not "smaller" than others, but collections are, and if numbers are conceived as collections of objects, then via the metaphorical mapping the properties of "smaller than" or "greater than" apply to them). Mathematical ideas, therefore, are offshoots of the same conceptual metaphorical system that manifests itself in everyday language and allows for precise forms of human imagination.

To grasp this more concretely, let's consider algebra problems constructed on the basis of two concepts of *time*, revealed by the differential use of *since* and *for* in common sentences such as the following (Danesi 2008):

> I have been living here *since* 2000.
> I have known Lucy *since* November.
> I have not been able to sleep *since* Monday.
> I have been living here *for* fifteen years.
> I have known Lucy for *nine* months.
> I have not been able to sleep *for* five days.

An analysis of the complements that come after *since* shows that they are metaphorical vehicles conceptualized as *points in time*, that is, they are grammatical forms that reflect a conception of *time* as a *point* on a *timeline* which shows specific years, months, etc.: *2000, November, Monday*. In contrast, the complements that follow *for* are vehicles that reflect a conception of *time as a quantity: fifteen years, nine months, five days*. These two concepts – *time is a point* and *time is a quantity* – are examples of the *conceptual blends* of which Lakoff speaks.

Now, these two concepts are found in the articulation of typical high school algebra problems, such as the following:

> Alex is five years older than Sarah. Four years from now Alex will be twice Sarah's age. How old is each one?

When the conceptual structure of the language in which the problem is cast is deciphered, solving it is a straightforward process. If x is used to represent Sarah's present age, then Alex's present age would be represented by $x + 5$. This is so, because *time* is a *point* on a *timeline* diagram on which Alex's age-point is "5 points" to the right of Sarah's age-point. Representing their two ages "four years from now" entails moving the two age-points "to the right by four" on the timeline. This translates into the algebraic expression $x + 4$ for Sarah and

$x + 5 + 4 (= x + 9)$ for Alex. Finally, to set up an algebraic relation between the two ages, it is necessary to enlist the *time is a quantity* metaphor. Alex's age is quantifiable as twice that of Sara's age. This is, metaphorically speaking, a difference in the size of the *containers* that hold the ages of each one – the container is a basic image schema. The final step is a straightforward one – namely, expressing in algebraic symbols the overall "metaphorical relation". We know that Alex's age four years from now is $(x + 9)$. We also know that this is *twice* Sarah's age at that time, i.e., $(x + 4) + (x + 4)$, or $2x + 8$. This leads to the equation: $x + 9 = 2x + 8$. Solving the equation reveals that Sarah is one year old and Alex six.

This shows several crucial things. First and foremost, it reveals that problems cast in language use the conceptual metaphorical system of the language in which they are cast. This in itself implies a close relation between language and mathematical thought. Second, it shows that mathematical ideas are likely unthinkable without some form of metaphorical interpretation. As mathematician Freeman Dyson has recently asserted, mathematicians are slowly coming to the realization that mathematics is, in a basic way, a metaphorical modeling system:

> Mathematics as Metaphor is a good slogan for birds. It means that the deepest concepts in mathematics are those which link our world of ideas with another. In the seventeenth century Descartes linked disparate worlds of algebra and geometry, with his concept of coordinates. Newton linked the worlds of geometry and dynamics, with his concept of fluxions, nowadays called calculus. In the nineteenth century Boole linked the worlds of logic and algebra, with his concept of symbolic logic, and Riemann linked the worlds of geometry and analysis with his concept of Riemann surfaces. Coordinates, fluxions, symbolic logic, and Riemann surfaces are all metaphors, extending the meanings of words from familiar to unfamiliar contexts. Manin sees the future of mathematics as an exploration of metaphors that are already visible but not yet understood. (Dyson cited in Marcus 2012: 89)

Now, the question becomes: Does the discovery of mathematical principles and concepts involve the same kind of thinking, but in reverse? That is to say, is mathematics produced through metaphorical reasoning? In MST terms, this would come as no surprise because, as mentioned briefly in the opening chapter, metaphor is a connective form (Sebeok, Danesi 2000). Science involves things we cannot see, hear, touch – atoms, gravitational forces, magnetic fields, etc. So, scientists use their imagination and their capacity to connect experiences, facts, interpretations, and so on to explore this invisible matter. The result is metaphor. Mathematical models, in this view, are connective forms becoming interpretants of given information and, thus, leading to further connections and insights. Marcus (2012: 184) insightfully writes on this theme as follows:

When mathematics is involved in a cognitive modeling process, both ana-
logical and indexical operations are used. But the conflict is unavoidable, because the model M of a situation A should be concomitantly as near as possible to A (to increase the chance of the statements about M to be relevant
for A too), but, on the other hand, M should be as far as possible from A so
as to increase the chance of M being investigated by some method which is
not compatible with the nature of A. A similar situation occurs with cognitive mathematical metaphors. Starting as cognitive model or metaphor for a
definite, specific situation, M acquires an autonomous status and it is open
to become a model or a metaphor for another, sometimes completely different situation. M may acquire some interpretation, but it can also abandon
it, to acquire another one. No mathematical *construction* can be constrained
to have a unique interpretation, its semantic freedom is infinite, because it
belongs to a fictional universe: mathematics. Mathematics has a strong impact on real life and the real world has a strong impact on mathematics, but
all these need a mediation process: the replacement of the real universe by
a fictional one.

And, as Lindsey Shorser (2012: 296) asserts, the embedded metaphorical
structure in a mathematical model is, *ipso facto*, its meaning: "In the absence
of sensory data, we perceive mathematical objects through cognitive metaphor, imbuing an abstract mathematical object with meaning derived from
physical experience or from other mathematical objects, ultimately linking every chain of metaphors back to concepts that are directly based upon physical
perceptions."

4.4. Blends

The use of metaphor to reason about mathematics and science now comes
under the cognitive science rubric of "embodied cognition" (see, for example,
Núñez et al. 1999; Lakoff 2008). In effect, mathematics is not "out there" to be
discovered by human minds; it is "inside" the body-mind complex, with the
physical and social context playing a determining role in how models are ultimately developed and how they interconnect to different domains of inquiry
(Bing, Redish 2007).

Although all these ideas are now exciting ones within cognitive science,
they constitute nothing new in MST, which has always seen metaphorical language as fundamental to model-building. But in mathematics it is becoming
a prominent factor in guiding research and the inquiry into learning. In one
study, Richard Lesh and Guershon Harel (2003) allowed students to develop
their own models of a problem space or situation, guided by connective forms
of instruction. The results were positive, leading to a whole spate of subsequent
studies confirming the findings. This kind of research is leading to the notion

that different problems in mathematics require different models and modeling systems – a fact that has been the thematic thread woven throughout this book. The many proofs of the Pythagorean theorem provide concrete evidence of this. There is no one proof, but many, depending on who, where, and why develops the proof. Nonetheless, the basic constituents of a proof will not change; the details will. The latter are due to context (historical, interpretive, social, and so on). As Harel and Larry Sowder (2007) have argued there exists a taxonomy of "proof schemes", which is based on the influence of convention vis-à-vis how proofs are modeled and how they are believed.

Lakoff and Núñez (2000) found that there are two basic conceptual metaphors within mathematical modeling systems, which they call *grounding* and *linking* metaphors. Grounding metaphors encode objects, directly grounded in experience. For example, one's perception of addition usually develops from adding objects to a collection. Linking metaphors connect concepts within mathematics itself that may or may not be based on physical experiences. Some examples of this in mathematics are numbers on a line, or understanding inequalities and absolute values properties within an epsilon-delta proof of limit. Linking metaphors also are involved in concepts such as negative numbers, which emerge from a connective form of artifactual modeling. As James C. Alexander (2012: 28) has argued, the introduction of negative numbers was the result of this kind of process, arising from an implicit oppositional model: *plus-versus-minus*:

> So what is it we have done? We started with the natural numbers, which do not always permit subtraction. Using the natural numbers, we made a much bigger set, way too big in fact. So we judiciously collapsed the bigger set down. In this way, we collapse down to our original set of natural numbers, but we also picked up a whole new set of numbers, which we call the negative numbers, along with arithmetic operations, addition, multiplication, subtraction. And there is our payoff. With negative numbers, subtraction is always possible. This is but one example, but in it we can see a larger, and quite important, issue of cognition. The larger set of numbers, positive and negative, is a cognitive blend in mathematics. For such a blend, there are two opposing aspects. On the one hand, there is nothing new. The new negative numbers "are" pairs of natural numbers, and all arithmetic is in terms of the natural numbers. There is no magic, and no question, philosophical or otherwise, as to whether the new negative numbers are really numbers. It is all a matter of definition and nomenclature. The new is included in the old. On the other hand, everything is new. The numbers, now enlarged to include negative numbers, become an entity with its own identity. The collapse in notation reflects this. One quickly abandons the (minuend, subtrahend) formulation, so that rather than (6, 8) one uses −2. This is an essential feature of a cognitive blend; something new has emerged.

Mark Turner (2012) refers to such knowledge-making as a *packing*-versus-*unpacking* process. One of the characteristics of mathematical representation is its tendency to compress information into compact and highly usable forms. When ideas are represented in this way, their structure becomes evident, and new ideas are possible because of the simplification afforded by the compression and abstraction. It is this hidden structure "packed into" a model that is the source of discovery in mathematics, as we have often remarked in this book and which we will discuss further in the next chapter.

In the view of Walter Whiteley (2012: 281) the sum and substance of mathematical modeling is just this form of packing ideas into a model or mathematical blend (as it has come to be known):

> Mathematical modeling can be viewed as a careful, and rich, double (or multiple) blend of two (or more) significant spaces. In general, modeling involves at least one space that is tangible – accessible to the senses, coming with some associated meaning (semiotics) – and a question to be answered! The mental space for this physical problem contains some features and properties that, if projected, will support reasoning in the blend. The mental space also includes a number of features and properties that will, if projected by distracting. Worse, these "irrelevant" features may suggest alternative blends that are not generative of solutions to the problem. Selective "forgetting" has been recognized as a crucial skill in modeling with mathematics – sometimes referred to as a form of abstraction.

The reason why blends work is because they are models, highlighting parts in relations that resemble relations among the parts of referents in the world. It is for this reason that Peirce called his graphs "moving pictures of thought" (CP 4.8–11).

4.5. Connective modeling

The question of blends leads to a discussion of connective modeling and the type of model called *metaform* by Sebeok and Danesi (2000). Essentially, a metaform is a model that connects forms in various domains either naturally or artifactually (or both) to produce an overall model – hence *metaform*. To differentiate it from the concept of *blend*, a metaform can be defined as the actual model itself produced by connective thinking and the term blend can be defined more specifically here as the neurological process involved in producing connective models. Metaforms are powerful models because they suggest change through their connectivity. Lyall Watson (1990: 42–43) puts it as follows (although he uses the traditional term *metaphor* rather than *metaform*):

> Metaphors are living things. They are slices of truth, evidence of the human ability to visualize the universe as a coherent organism. Proof of our capac-

ity, not just to see one thing in another – as Blake saw the world in a grain of sand – but to change the very nature of things. When a metaphor is accepted as fact, it enters mythology, but it can also take on an existence in the real world.

The metaforms of Euclidean geometry, for instance, gave the world a certain kind of visual structure for millennia – a world of relations among points, lines, circles, etc. A new metaform of this structure emerged when Nicholas Lobachevski literally imagined that Euclid's parallel lines would "meet" in some context, such as at the poles of a globe, thus giving the visual world a different structure.

At this point it is necessary to elaborate upon the concept of connective modeling, since it can be easily (yet not quite correctly) related to the theory *associationism* in psychology and linguistics. This is the theory that the mind comes to form concepts by combining simple, irreducible elements through mental connection. One of the first to utilize the notion of association was Aristotle, who identified four strategies by which associations are forged: by similarity (an orange and a lemon), difference (hot and cold), contiguity in time (sunrise and a rooster's crow), and contiguity in space (a cup and saucer). British empiricist philosophers John Locke (1975[1690]) and David Hume (1902[1749]) saw sensory perception as the underlying factor in guiding the associative process; that is, things that are perceived to be similar or contiguous in time or space are associated to each other; those that are not are kept distinct. In the nineteenth century, the early psychologists, guided by the principles enunciated by James Mill in his *Analysis of the Phenomena of the Human Mind* (1829), studied experimentally how subjects made associations. In addition to Aristotle's original four strategies, they found that factors such as intensity, inseparability, and repetition play a role in associative processes: for example, arms are associated with bodies because they are inseparable from them; rainbows are associated with rain because of repeated observations of the two as co-occurring phenomena; etc.

Associationism took a different route when Ivan Pavlov (1902) published his famous experiments with dogs. As is well known, when Pavlov presented a meat stimulus to a hungry dog, the animal would salivate spontaneously, as expected. He termed this the dog's "unconditioned response" – an instinctual response programmed into each species by nature. After Pavlov rang a bell while presenting the meat stimulus a number of times, he found that the dog would eventually salivate only to the ringing bell, without the meat stimulus. The ringing by itself, which would not have triggered the salivation initially, had brought about a "conditioned response" in the dog. It was thus by repeated association of the bell with the meat stimulus that the dog had learned something new – something not based on instinctual understanding. The Pavlovian theory of conditioning is useful on various counts, despite the many questions it raises and the debates it has engendered. It simply confirms that forms (bell ringing) and displaced referents (meat) are blended together in the minds of various species. The difference in the associative powers of humans is the fact that the blending process is not restricted to conditioning but entails creative thinking. So, rather than describe metaforms as the result of associations in the traditional psychological senses of that word, it is more appropriate to describe them as the result of creative connective processes derived from the use of a specific form of imaginative thought that can be called *sense implication* (Danesi 2004a). Sense implication is the process of sensing that things are connected (in all the polysemous meanings of the word *sense*). The connection may cross modal boundaries. It connects existing or new forms, varying in the type and degree of the implication. It can also be guided or constrained by cultural practices and traditions and thus dictated by historical practices rather than strict sensory entailments.

The great American philosopher Susanne Langer (1948: 129) compared sense implication (which of course she does not name in this way) appropriately, to a "fantasy:"

> Suppose a person sees, for the first time in his life, a train arriving at a station. He probably carries away what we should call a "general impression" of noise and mass. Very possibly he has not noticed the wheels going round, but only the rods moving like a runner's knees. He does not instantly distinguish smoke from steam, nor the hissing from the squeaking. Yet the next time he watches a train pull in the process is familiar. His mind retains a fantasy which "means" the general concept, "a train arriving at a station". Everything that happens the second time is, to him like or unlike the first time. The fantasy was abstracted from the very first instance, and made the later ones "familiar."

Metaforms result not only from metaphor but other connective processes, such as metonymy (as Lakoff and Núñez also assert). There is no need to bring up the differential functions and forms between these two processes here. Suffice it to say that, unlike metaphor, metonymy does not function to create knowledge through connective reasoning, but rather it allows us to cast specific kinds of light on certain situations, so as to be able to make some comment on them. It is an indexical modeling process that allows relations to be made explicit. Marcus (2012: 146) makes the appropriate following observation in this regard:

> Complementary to metaphorical thinking is metonymical thinking. The former is related to iconic thinking, the latter, to indexical thinking. Metonymy is everywhere in mathematics, either as *pars pro toto* or as in *if-then* thinking. The whole mathematical enterprise is metonymical, since mathematics is looking for a suitable representation of infinity by countable forms, then to reduce the latter to a finite representation and after that to reduce the large finite to the small finite. There is the claim that mathematics is the science of approximations; but approximation is a metonymical notion. Most real numbers have essentially infinite representations (decimal or by continuous

fractions) and we try to capture finite parts of them, as large as possible. This process never stops. A famous example is the constant effort to capture the decimals of the number p. This began with Archimedes and is continued today in computer programs and by clever procedures such as those found in the notebooks of Ramanujan. The basic metonymy, *if-then*, is at the root of the deductive thinking essential in the final presentation of mathematical proofs. It is the main tool to validate a mathematical theorem.

In a fundamental way, connective reasoning and modeling undergirds all modeling in mathematics, at least in a generic metaformal way. This is why one model leads to another and then to another and so on. Recall Pascal's triangle. It is a metaform that results from a layout of the expressions in the binomial expansion, $(a + b)^n$, revealing how certain numbers are related to others, as well as how shapes and number patterns are intertwined. The numerical coefficients in the expansion form the shape of a triangle with infinite dimensions. This connection between a geometrical form, an algebraic concept, and the numerical coefficients involves metaformal thinking. It is by noticing the numbers in the triangular metaform that a further pattern emerges: any number in a row is produced by the sum of the two numbers above it. Now, Pascal's triangle crops up in many areas of mathematics. It turns up, as we saw, in the calculation of probabilities. It also turns up in combinatorial analysis. Indeed, many historians claim that without it these would never have occurred to mathematicians. It also crops up in curves.

The same story can be repeated over and over for all kinds of mathematical metaforms. These are externalizations of blends in the brain, which often occur unconsciously. Mathematicians are sometimes so inspired by a problem that, as Stanislas Dehaene (1997: 151) puts it, "truth descends on them", leading them to imagine some new hypothesis, conjecture, or theory. The most famous example of this is provided by the French mathematician Pierre de Fermat (1601–1665), as briefly noted previously. While reading Diophantus's *Arithmetica*, Fermat became keenly interested in Pythagorean triples – recall that these are sets of three numbers, a, b, and c, for which the equation $a^2 + b^2 = c^2$ is true. These numbers had captivated the imagination of Diophantus, who included a long insightful discussion of them in his book. That discussion stimulated Fermat's own imagination. In the margin of his copy of the book, he wrote the following enigmatic words (cited by Pappas 1991: 150):

> To divide a cube into two cubes, a fourth power, or in general any power whatever above the second, into two powers of the same denomination, is impossible, and I have assuredly found an admirable proof of this, but the margin is too narrow to hold it.

For more than 350 years, mathematicians across the world were intrigued by Fermat's claim, trying valiantly to come up with a proof, but always to no avail, although a number of special cases were settled. Gauss proved that $a^3 + b^3 = c^3$ had no positive solutions, and Fermat himself proved the untenability of $a^4 + b^4 = c^4$. The French mathematician Adrien Marie Legendre (1752–1833) gave a proof that $a^5 + b^5 = c^5$ had no solutions. And Lejeune Dirichlet (1805–1859) showed that $a^{14} + b^{14} = c^{14}$ had no solutions. But no general proof, as Fermat envisioned it, was discovered until in June 1993 Andrew Wiles, an English mathematician teaching at Princeton University, declared that he had finally proved what had come to be known as Fermat's last theorem. Wiles's proof is the result of connecting and modifying previous metaforms used to explore the theorem. Two previous ideas, in fact, were crucial to Wiles's proof: the elliptic curve and the modular form. An elliptic curve is a curve of the form expressed by the following equation:

$$y^2 = x^3 + rx^2 + sx + t \text{ for integers r, s, and t}$$

A modular form is a formula generalizing the Möbius transformation:

$$f(z) = (az + b)/(cz + d)$$

Two mathematicians, Yutaka Taniyama and Goro Shimura, hypothesized that every elliptic curve is associated with a modular form. It was quickly seen that a proof of the Taniyama-Shimura conjecture would imply the truth of Fermat's last theorem. In 1984 another mathematician, Gerhard Frey, saw that if Fermat's theorem were false – so the equation $a^p + b^p = c^p$ held for some positive integers a, b, c, and prime number p – then the elliptic curve $y^2 = x^3 + (b^p - a^p) x^2 - a^p b^p x$ would have such weird properties that it could not be modular. This would then contradict the Taniyama-Shimura conjecture. Wiles took Frey's idea and set out to prove the special case of the Taniyama-Shimura conjecture that implied the truth of Fermat's last theorem. Andrew Wiles, together with another English mathematician, Richard L. Taylor, published his proof in May 1995 (Wiles, Taylor 1995).

Fermat's last theorem still haunts some mathematicians, for the simple reason that the Wiles-Taylor proof was certainly not what Fermat envisioned. Their proof required a computer program and depended on mathematical work subsequent to Fermat. In a pure sense, therefore, the Wiles-Taylor proof does not really constitute a historically appropriate resolution to Fermat's last theorem. As Marcus (2012: 112) has recently written, the use of computer programs to carry out proofs is problematic, yet useful, for the epistemology of traditional mathematics:

> The use of some computer programs as components of a mathematical proof involves delicate problems of a meta-theoretic nature, related to the situation whereby a computer program is checked by another computer program. On

the other hand, the new situation, in strong contrast with a long tradition, that a mathematical proof could have, besides its logical component, an empirical-experimental-physical component, raises many questions, that we will have to face in the next decades. If twenty years ago the opposition to computer-aided proofs was strong among mathematicians, in the meantime this opposition decreased in intensity and there are voices according to which such proofs may have sometimes an advantage with respect to traditional proofs. They could be more illuminating, by pointing out better the structure of the proof. There is also the fact that such long proofs involve often a very large number of authors, each of them reaching the understanding of a small part of the proof, but (almost) none of them having its global understanding. The typical example in this respect is the proof of the theorem of classification of finite simple groups, having hundreds of authors.

Fermat left behind a true mathematical detective mystery. What possible "simple proof" could he have been thinking of as he read Diophantus's *Arithmetica*? There are undoubtedly some mathematicians still trying to find Fermat's mysterious proof, if indeed he ever had one in the first place. As Ian Stewart (1987: 48) aptly puts it, "Either Fermat was mistaken, or his idea was different". One line of inquiry that has been pursued in investigating Fermat's last theorem is in the domain of equation systems and of their reference to dimensions in terms of exponents. As Diophantus asserted, the Pythagorean relation holds because it links one number (the square of the hypotenuse) to two others (the squares of the other two sides). This relation is reified by a figure in two-dimensional space, namely, a right-angled triangle: $a^2 + b^2 = c^2$. The c^2 in this equation can be called the *Pythagorean term*, for the sake of convenience, and represented with p^n. In the above case $n = 2$, which of course indicates the number of dimensions to which the Pythagorean relation refers. It would therefore seem logical to assume that in three-dimensional space, the corresponding relation would have to take into account this new dimension. This could thus be represented by one more variable in the left side of the equation, corresponding to the number of dimensions, $n = 3$:

$$a^3 + b^3 + c^3 = p^3$$

Following this line of reasoning, in n-dimensional space the Pythagorean relation could be expressed as follows:

$$a^n + b^n + c^n + d^n + \ldots n^n = p^n$$

This is an equational metaform showing that a variable is added to the equation in accordance with the number of dimensions. In a seven-dimensional space, where $n = 7$, there would therefore be seven variables to the left side of the equation:

$$a^7 + b^7 + c^7 + d^7 + e^7 + f^7 + g^7 = p^7$$

It is certainly beyond the scope of the present discussion to go into the merits or demerits of this proposal and the kinds of investigations it might entail. What kind of seven-dimensional figure would constitute the reification of the above formula? Could Fermat's proof have been consistent with this line of inquiry? In brief, this episode in the history of mathematics shows how much discovery is the result of connective modeling.

As another example, consider the discovery of π in mathematics. As mentioned previously, π appears in many other domains, in equations and area formulas of all kinds. However, π appears outside of mathematics. Consider the following truly incredible appearance of this constant. Take a piece of cardboard and a needle, marking parallel lines on the cardboard as shown below, spacing them twice the length of the needle apart, and tossing the needle in the air so that it falls on the cardboard:

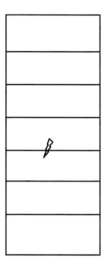

The object of the experiment is to determine the relation between the number of tosses of the needle and the number of times the needle touches a line on the cardboard. It has been found that as the number of tosses of the needle increases, the ratio of the number of tosses to the number of times the needle touches the line approaches π. Why this is so defies reason. Where is the circle figure in this and in other manifestations of π? As Robert B. Banks (1999: 57) observes, it boggles the mind to think that π "appears in a large number of mathematical problems that have nothing to do with circles". One of these serendipitous appearances is also in the so-called *Basel problem*, which was to find the sum of the reciprocals of perfect squares:

1 + 1/4 + 1/9 + 1/16 + 1/25 …

The answer turns out to be around 1.645, but no one could calculate its exact value. It was Euler who showed that the sum is $\pi^2/6$. He then ingeniously calculated the sum of the reciprocals of all the 4th powers ($\pi^4/90$), the 6th powers ($\pi^6/945$), and so on right to the 26th powers. It is mind-boggling to realize that π shows up in this calculation, as if by magic. And, really, no one knows why. But the story does not end there. Euler generalized his finding with his *zeta function* z(k) as follows:

$$z(k) = 1 + (1/2)^k + (1/3)^k + (1/4)^k + (1/5)^k \ldots$$

so

$$z(2) = \pi^2/6$$
$$z(4) = \pi^4/90$$
$$z(6) = \pi^6/945$$

and so on

It turns out that z(k) is defined for every real number k > 1. But the harmonic series 1 + 1/2 + 1/3 + 1/4 + 1/5... has, of course, no finite sum, so z(1) is not defined. Can we define the zeta function for other numbers? It was Bernhard Riemann who discovered a way of doing so for every number, real or complex. His function is now known as the *Riemann zeta function*. It turns out that problems involving prime numbers are connected with the zeroes of the *zeta function*, the solutions of z(z) = 0 in the complex plane. It also turns out that the function has zeros at –2, –4, –6, –8,...and that all other zeros occur within a vertical strip between 0 and 1, called the *critical strip*. All known zeros in the strip occur at points of the form 1/2 + ki. So, they lie on a *critical line*, as it is called. Now comes a truly fascinating question: do all the zeros in the critical strip lie on this line? This has come to be known as the *Riemann hypothesis*. No one has proved this yet.

Such crystallizations of π within mathematics and the extensions of these into other domains, such as prime number theory, are amazing in themselves. What is even more astonishing is that π also appears in formulas describing natural formations and phenomena, such as rivers. The actual length of a river winding from its source to where it enters the sea is a much longer distance than a straight line drawn between the source and end points. As rivers cut through paths according to land formation, cutting through soft rock, flowing downhill, and so on, it has been shown that if we divide the length of the winding path by the length of the hypothetical straight path, the answer is always very close to π. Many question are raised by such serendipities. Is π a natural model of something intrinsic in the world? But how could this be, since it is a model of mathematical pattern derived artifactually by measuring circles?

In a fascinating 1997 movie, titled π: *Faith in Chaos*, by American director Darren Aronofsky, a brilliant mathematician, Maximilian Cohen, teeters on the brink of insanity as he searches for an elusive pattern or code hidden in π. For the previous ten years, Cohen has been on the verge of his most important discovery, attempting to decode the numerical pattern beneath the chaotic stock market. As he is on the threshold of a solution, real chaos is swallowing the world in which he lives. Pursued by an aggressive Wall Street firm set on financial domination and a Kabbalah sect intent on unlocking the secrets hidden in their ancient holy texts, Cohen races to crack the code, hoping to defy the madness that looms before him. Instead, he uncovers a secret for which everyone is willing to kill him, and he succumbs to the madness in the end (figuratively speaking). The movie suggests that mathematical concepts, derived artifactually, may hide the truths about the world. Perhaps the human brain is a simple part of the world and gathers information from it unconsciously, turning it into signs, sign systems, and modeling systems.

The same kind of connective modeling outcomes can be seen in the history of the golden section or ratio (Dunlap 1997; Livio 2002). This results from the division of a line segment in such a way that the ratio of the whole segment to the larger part is equal to the ratio of the larger part to the smaller part. The ratio, represented with the Greek letter *phi* (ϕ) is approximately 1.61803 to 1:

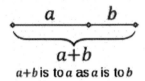

a+b is to *a* as *a* is to *b*

As is well known, ϕ is $(1 + \sqrt{5})/2$, a number mentioned at the beginning of Book VI of Euclid's *Elements*. Since antiquity many philosophers, artists, and mathematicians have been intrigued by the golden ratio, which Renaissance writers called the *divine proportion*. It is widely accepted that a figure or form incorporating this ratio exhibits a special beauty. Incidentally, the Egyptians also used the golden ratio, calling it the "sacred ratio", in building the Great Pyramid at Giza. Artists and architects throughout history have employed this ratio. The rectangular face of the front of the Parthenon in Athens has sides whose ratio is golden. The ratio of the height of the United Nations Building in New York (built in 1952) to the length of its base is also golden. Remarkably, the ratio of two successive terms in the Fibonacci sequence converges on the golden ratio: for example, 5/8 = .625, 8/13 = .615, 13/21 = .619, etc. It has also been found in nature, such as in the curve of a nautilus shell. One cannot help but be struck by such serendipities. As Pappas (1999: 5) appropriately puts it, this ratio mysteriously "sets off a truly amazing chain of mathematical interconnections."

One of the more fascinating manifestations of the ratio is in the golden rectangle.

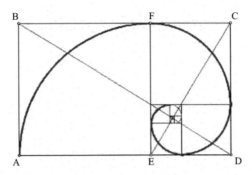

First, the method of construction of this rectangle is itself interesting. Without going into details here, one can continue to draw rectangles within it as shown. If we draw a spiral through the corners of the golden rectangles, the resulting spiral is the same one found in sunflower heads, pineapples, pinecones, nautilus shells, and many other natural formations. It would seem that a form derived by a mathematical rule and based on a mathematical metaform, f, occurs throughout nature. Which came first, one can ask? The power of mathematics to describe natural phenomena has led many to wonder if everything is determined in the world, and all we have to do is uncover the patterns in it. To quote Pierre-Simon Laplace (1749–1827), a leading eighteenth-century French mathematician, from the introduction of his *Essai philosophique sur les probabilities*:

> We may regard the present state of the universe as the effect of its past and the cause of its future. An intellect which at a certain moment would know all forces that set nature in motion, and all positions of all items of which nature is composed, if this intellect were also vast enough to submit these data to analysis, it would embrace in a single formula the movements of the greatest bodies of the universe and those of the tiniest atom; for such an intellect nothing would be uncertain and the future just like the past would be present before its eyes. (Laplace cited in Flood and Wilson 2011: 121)

This is, of course, a pipe dream, as Davis and Hersh (1986) have cogently argued. Certainly, there is no doubt serendipities between mathematical metaforms, nature, and human constructions lead to feelings of awe and consternation. The discovery of related patterns is the result of connective modeling. Perhaps connective reasoning results from nothing more simple than experience of the world. Here's a banal, yet instructive, example of this. Engineers sometimes use the triangle form to make bridges stronger. The reason is that four-sided shapes, such as squares and rectangles, can be bent and deformed by forces such as the wind:

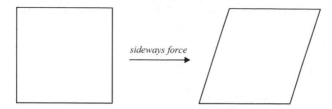

However, by adding another beam across the middle of a square, it is turned into two triangles, making it much stronger, since the sides cannot be moved without breaking the beam.

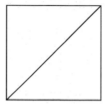

In effect, by studying geometrical forms and relating them to physical forms, engineers develop an insight that will suggest to them how to make bridges stronger. The same kind of account could be given for virtually all architectural forms. Domes, for example, act like arches, which are geometrically strong shapes. In building the dome for the great cathedral in Florence in the Renaissance, the great Filippo Brunelleschi arranged the bricks so that the stresses were transferred sideways to the wooden struts in the arch form. Any other arrangement would have allowed the force of gravity to act downwards, forcing the bricks to fall down. Empirical science works in this way, as Galileo's early experiments showed – experiments that utilized traditional arithmetical and geometrical concepts in aid of the discovery of natural principles. Among the most important results of this search between natural and artifactual models were the law of the pendulum and the law of freely falling bodies. Galileo observed that pendulums of equal length swing at the same rate whether their arcs are large or small. Galileo's law of falling bodies states that all objects fall at the same speed, regardless of their mass; and that, as they fall, the speed of their descent increases uniformly.

As one final example, consider the emergence of fractal geometry. This branch is concerned with complex shapes called fractals, which consist of small-scale and large-scale structures that resemble one another. Certain fractals are also similar to natural objects, such as coastlines or branching trees. Although fractals seem irregular, they have a simple organizing principle. The origins of fractal geometry can be traced to the work of Bolzano and Poincaré, but the topic was greatly developed by mathematician Benoit Mandelbrot (1977), who found that random fluctuations in nature and in human affairs form geometrical patterns, which he called *fractals*, when they are reduced to smaller elements. A fractal is defined as any form that is altered by application of a transformational rule to it *ad infinitum*. Coined from the Latin word *fractus*, Mandelbrot's term suggests fragmented, broken and discontinuous phenomena. But, as it turns out, fractals disclose a strange type of hidden pattern in shapes that would otherwise appear random to the naked eye. Here's an example of a fractal form, known as the *Sierpinski triangle*, named after the Polish mathematician Vaclav Sierpinski (Barnsley 1988). As can be seen, it is produced by repeating the triangle form over and over:

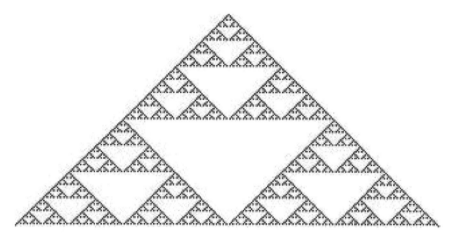

Remarkably, snowflakes have a fractal form. Like the Sierpinski triangle, a snowflake seems to be generated by nature with a transformational rule – a connection discovered by Swedish mathematician Helge von Koch (1870–1924) in 1904. To construct it, Koch took an equilateral triangle, replacing the middle third of each side by two other sides of the triangle, giving a peak on each side of the original triangle. Repeating this process produced a Koch snowflake, which is incredibly similar to real snowflakes:

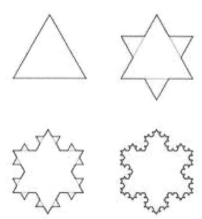

Fractal shapes were known to the human imagination long before fractal geometry provided a theory for them. They turn up in the ancient world, in art, and in certain artifacts. An early prototype of the Sierpinski figure crops up in a thirteenth-century pulpit in the Ravello cathedral in Italy, designed by Nicola di Bartolomeo of Foggia. In Mahayana Buddhism, the fractal nature of reality is captured in the Avatamsaka Sutra by the god Indra's net, a vast network of precious gems hanging over Indra's palace, so arranged that all the gems are reflected in each one. In recent times, artists Dalí, Pollock, and Escher have exploited the fractal technique of creating a new shape out of repeated copies of another. Fractals are perfect models for understanding the shapes of such natural formations as fern leaves, lava flows, coastlines, and mountain terrains (Barnsley 1988; Devlin 1997)

Fractal geometry is a connective modeling system in its own right, grafting ideas from several fields of mathematics. Marcus (2012: 179) makes the following appropriate observation on this remarkable system, which interconnects not only with other domains of mathematics, but also with art and poetry:

What art and poetry anticipated in the 19th century, together with some phenomena pointed out by Weierstrass, Peano and Koch, related to curves devoid of tangents in all their points, became explicit in the mathematics of the second half of the past century, when Benoit Mandelbrot invented *the fractal geometry of nature*. Its idea is that nature, in most of its aspects, is not at all simple and regular. Clouds, ocean coasts, Brownian motion, snowflakes, mountains, rivers don't fit with the regular objects of traditional geometry. Even celestial bodies, longtime considered models of regularity, prove to be less regular than they were supposed to be. How to approach this world of high complexity? The answer proposed by Mandelbrot is the notion of a fractal object. Such objects are obtained as limits of some asymptotic processes, starting with some regular figures. If the first steps of these processes are visible objects and fit with the simplicity of traditional geometry, as soon as we go to next steps the new objects become less and less visible and regular. At the limit, we get completely invisible, however perfectly intelligible objects, the fractal ones. What makes them very attractive is their inner, hidden simplicity, in contrast with their outer complexity: in a fractal object, there is a remarkable phenomenon of self-similarity: it repeats at its different levels in the same structure As a matter of fact, everybody can test this fact looking carefully at the structure of a tree in the forest.

5. Mathematical discovery

5.1. Introduction

Discovery in mathematics is guided, as we have seen throughout this book, by modeling – one model (natural or artifactual) leading to another, and then to another, and so on. As the modeling process becomes more and more connective, it leads to the opening up of new vistas, new branches, new conceptualizations within the discipline. It is for this reason that we can call mathematics itself a modeling system, or more accurately a meta-modeling system.

Discovery of ideas comes about in various ways, as we have seen throughout the previous discussion, not just connective reasoning. One of these is through the modification of an existing model or through the compression of some model or formula, as the blending theorists have recently started to argue (see Chapter 4). Essentially, if you change the form of something in some way it becomes itself a new model that suggests other meanings and ideas. A simple illustration of this will bring this out. Consider multiplying the number 6 sixteen times in a row:

$$6 \times 6 \times 6 \times 6 \times 6 \times 6 \times 6 \times 6 \times 6 \times 6 \times 6 \times 6 \times 6 \times 6 \times 6 \times 6 = \ ?$$

As it stands, the computational task presents itself as an unwieldy one, not only to carry out physically, but also just to read on the page. To make it easier to do the latter mathematicians came up with an abbreviation sign called the *exponent*, which is a superscript number or letter written over a base (the number that is to be multiplied by itself a certain number of times). The idea of using such a simple notation surfaced as early as the fourteenth century, but it entered general use through the influence of Descartes in the seventeenth century. With this notation, the above multiplication can be compressed into the simple form 6^{16}. Although the work of carrying out the computation is still problematic, the ease with reading the number of factors involved has been simplified noticeably.

Now, as mathematicians began using this notation they started discovering new and unexpected things, such as, for example, that n° is equal to 1, thus producing a new meaning to the uses of zero. The new notation laid the basis for the branch of mathematics known as logarithms. For example, in the exponential formula $2^3 = 8$, the exponent 3 is the logarithm of the number 8 to the base 2. This statement is written as $\log_2 8 = 3$. Logarithms can now be used to simplify computation tasks considerably as the following well-known illustrative case used in mathematics school books makes clear. Suppose one wants to calculate the number of one's ancestors in each of three previous generations. Each of us has 2 parents, so we have 2 ancestors in the first generation. This calculation can be expressed as $2^1 = 2$. Each of our parents has 2 parents, that is, $2 \times 2 = 2^2 = 4$ ancestors in the second generation. Each of our grandparents has 2 parents: $4 \times 2 = 2 \times 2 \times 2 = 2^3 = 8$ ancestors in the third generation. The calculation continues in this way. In which generation do we have 1,024 ancestors? We can translate this question into one involving exponents: For which exponent n is it true that $2^n = 1,024$? You can find the answer by multiplying 2 by itself until you reach 1,024. But if you know that $\log_2 1,024 = 10$, you know right away that the answer is 10.

It was John Napier (1550–1617), a Scottish mathematician, who published the first discussion and table of logarithms in 1614, although Jobst Burgi (1552–1632) of Switzerland independently discovered logarithms at about the same time. In the early 1600s, the Englishman Henry Briggs (1561–1630) introduced logarithms to the base 10 and Adriaen Vlacq (1600–1667) of the Netherlands completed Briggs's work. Today, computers have eliminated the need for logarithms for computation purposes. But the branch still exists as an important one in its own right and in its cross-over to other branches where logarithms appear as inherent forms and models of one kind or the other.

5.2. Structural economy

Mathematics is both discovered and invented. The line between the two is a blurry one. Certainly, as the episode of exponents and logarithms shows, invention is sometimes a precursor to discovery. Exponential notation had a simple purpose – to represent the multiplication of identical digits in a compressed easy-to-read way. But as mathematicians used the notation they started discovering new and unexpected things imprinted in it. This case in point brings out an intrinsic principle in the theory of representation and modeling that can be called *structural economy* (Danesi 2007: 34–45). This is the theory that human semiotic systems evolve in a compressed fashion in order to make cognitive tasks increasingly more efficient and economical. Exponential notation is structurally more economic than traditional multiplication layouts. The numerical systems used today (such as the decimal and binary ones) are structurally more economical than were older ones, which had abundant symbolic materials with which to represent numerical concepts (such as the Roman one). In the numerical systems we use today, representational length (the layout of numbers) has been greatly reduced.

Structural economy often correlates with frequency – the more frequently a form or category is used the more likely it is that it will be compressed. This is why common abbreviations and acronyms are used to spell frequently-used words: for example, *ad* (advertisement), *photo* (photograph), *NATO* (North Atlantic Treaty Organization), *laser* (light amplification by stimulated emission of radiation), *DVD* (digital video disc), *it's* (it is), *UN* (United Nations), *CD* (compact disc), and so on. These compressed forms save space and effort in writing. Structural economy is also a characteristic force in the use of tables, technical and scientific material, indexes, footnotes, and bibliographies.

Psychologically, manifestations of structural economy suggest the presence of a Principle of Least Effort (PLE) in human representational and modeling behavior, which can be defined as the propensity to reduce the physical structure of signs and sign systems so as to minimize effort on all counts (cognitive, interpretive, and so on). The PLE came to the forefront in the 1930s and 1940s, when the Harvard linguist George Kingsley Zipf published a series of key studies showing that the most frequently-used words in written texts (newspaper articles, novels, etc.) were also the shortest in length (in terms of the number of alphabet characters that constituted them). The higher the frequency of a word, the more likely it was, Zipf concluded, to show an economy of form (fewer letters, fewer syllables, etc.). More technically, Zipf discovered that there was an inverse correlation between the length of a specific word (in number of phonemes) and its rank order in the language (its position in order of its frequency of occurrence in texts of all kinds). The higher the rank order of a word (the more frequent it was in actual usage), the more it tended to be smaller phonemically. Articles, conjunctions, and other such forms, which had a high rank order, were typically monosyllabic, consisting of 1 to 3 phonemes (see, for example, Zipf 1935, 1949).

To grasp the substance and validity of Zipf's discovery, all one has to do is take the words in a substantial corpus of text, such as a novel or a newspaper, and count the number of times each word in it appears. Variants of the same basic word are to be counted as separate tokens of the word: for example, *play, played, playing* all count as three manifestations of the same word. If one were then to plot the frequencies of the words on a histogram according to their length, the curve would approach a line with a slope as shown below.

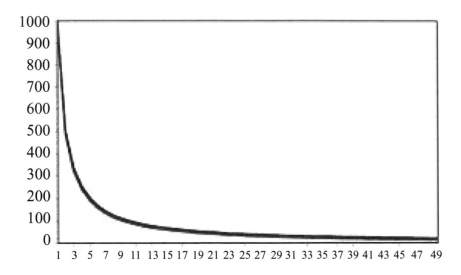

The same type of result emerges from counting words in textbooks, recipe collections, speeches, novels, and all other texts. The larger the corpus the more the curve tends towards this slope. Zipf found that this was true of any language, and not only words, but also grammatical systems, Chinese characters, and nonverbal forms and symbol systems, including mathematics (see, for example, Saichev et al. 2010). The gist of this research is that humans always seem to seek the path of least resistance or effort in their representational and communicative activities. What is startling is that, in so doing, they chance upon unexpected new pattern or meaning in the compressed forms, as we saw above with respect to exponential notation. Some scholars have rejected the PLE, suggesting that Zipf curves are simple consequences of his particular method of regarding a message source as a random process and, thus, a function of probability considerations, not psychological propensities. But many others have accepted the PLE as a general framework for explaining such everyday phenomena as abbreviation and acronymy in language. It certainly seems to apply to mathematical blends or metaforms, which, as we saw, compress information from various sources through connective reasoning, producing a model that contains the main elements of these forms in it.

For the sake of historical accuracy, it should be mentioned that some of Zipf's ideas were prefigured by various scientists in the nineteenth century. For example, the American astronomer Simon Newcomb found in 1881 that if the digits used for a task are not entirely random, but somehow socially based, the distribution of the first digit is not uniform – 1 tends to be the first digit in about 30% of cases, 2 will come up in about 18% of cases, 3 in 12%, 4 in 9%, 5 in 8%, etc. Newcomb came to this discovery by noticing that the first pages of books of logarithms were soiled much more than the remaining pages. In 1938, physicist

Frank Benford investigated listings of data, finding a similar pattern to the one uncovered by Newcomb in the area of income tax and population figures, as well as in the distribution of street addresses of people listed in phone books. Zipf's main contribution was in showing empirically that patterns of this type manifest themselves regularly and blindly in all kinds of human representational efforts, especially in language. Shortly after the publication of Zipf's ideas, Mandelbrot, the founder of fractal geometry (Chapter 4), became fascinated by them, detecting in them an indirect confirmation of what is called a scaling law in biology. Mandelbrot made significant mathematical modifications to Zipf's original law – modifications that are used to this day to study structural economy phenomena in language and other sign systems. The basic implication to be derived from Zipf's analysis can be reformulated in semiotic terms as the tendency in modeling towards the compression of information, while the meaning range of the signifying elements in the compression process is preserved and, in some cases, even expanded.

At first, Zipf did not take meaning into consideration. However, when he did, he also found some fascinating patterns. For example, he discovered that, by and large, the number of words (n) in a lexicon or text was inversely proportional to the square of their meanings (m): $(n)(m)^2 = C$. In 1958, psycholinguist Roger Brown subsequently claimed that Zipfian analysis applied to meaning in a general way because it correlated with the Whorfian concept of codability (Whorf 1956). This implies several corollaries. The most widely known one is that language users encode the concepts that they need and thus develop specialized vocabularies and forms to respond appropriately to their needs and environments. This also determines the size of their vocabularies and the constitution of their grammars. If speakers of a language need to use colors a lot, then they will develop more words for color concepts than do other languages that do not need them as much; if they have to deal with snow in their environment on a regular basis, then they will develop more words for types of snow than will other cultures where snow concepts are irrelevant. Codability also extends to the grammar of a language, which is seen as a culture-specific guide for understanding situational realities. Brown (1958b: 235) puts it as follows:

> Zipf's Law bears on Whorf's thesis. Suppose we generalize the finding beyond Zipf's formulation and propose that the length of a verbal expression *(codability)* provides an index of its frequency in speech, and that this, in turn, is an index of the frequency with which the relevant judgments of difference and equivalence are made. If this is true, it would follow that the Eskimo distinguishes his three kinds of snow more often than Americans do. Such conclusions are, of course, supported by extralinguistic cultural analysis, which reveals the importance of snow in the Eskimo life, of palm trees and parrots to Brazilian Indians, cattle to the Wintu, and automobiles to the American.

This interpretation of Zipfian theory has, as mentioned, been critiqued by a number of influential scholars (for example, Miller 1951, 1981; Lucy 1992). Miller (1981: 107) dismisses it as follows: "Zipf's Law was once thought to reflect some deep psychobiological principle peculiar to the human mind. It has since been proved, however, that completely random processes can also show this statistical regularity." But a resurgence of interest in Zipf's law suggests something very different – namely, that Zipf was onto something rather "deep" indeed, although some refinement or modification of this theory was needed. Empirical work by Ferrer i Cancho (Ferrer i Cancho, Sole 2001; Ferrer i Cancho 2005; Ferrer i Cancho et al. 2005), for instance, has shown that greater effort is required by speakers, since they have to make themselves understood; whereas listeners must work harder to ensure that they interpret messages correctly. In other words, Zipfian theory does not operate blindly but rather in response to communicative and other pragmatic factors. When there are small shifts in the effort expended by speaker or hearer, changes occur cumulatively because they alter the communicative and interpretive entropy of the whole system. Interestingly, Zipfian laws have been found in other species. For example, McCowan et al. (1999) discovered that Zipf's basic theory applies to dolphin communication.

As Colin Cherry (1966: 103) aptly observed, Zipf did indeed understand the relation between form and meaning, unlike what his critics believed:

> When we set about a task, organizing our thoughts and actions, directing our efforts toward some goal, we cannot always tell in advance what amount of work will actually accrue; we are unable therefore to minimize it, either unconsciously or by careful planning. At best we can predict the *total likely* work involved, as judged by our past experience. Our estimate of the "probable average rate of work required" is what Zipf means by *effort*, and it is this, he says, which we minimize.

As we have seen throughout this book, Zipfian analysis is extremely useful in explaining, at least in part, why compression is one of the factors that leads to discovery. But then this raises a larger question, which can be formulated simply as: Why should this be so? It also invokes the question of where the line between invention and discovery lies. Is the Pythagorean theorem invention or discovery? It is invention insofar as it is set up to describe patterns that occur in considering rope stretching that produce a specific triangular form; it is discovery in the fact that as a theorem it leads to an understanding of space that was not conscious before. Subsequently, it led to further discoveries and insights, as we saw.

The same type of explication can be given for virtually any mathematical discovery. Take, once again, Pythagoras' classification of numbers into *even* and *odd*. Although this seems to be simply a convenient way of describing numbers, as it turns out, the classificatory schema morphed into a model of hidden number properties such as, for example, the fact that prime numbers, other than 2, are all odd numbers. The whole branch of number theory would never have crystallized without a schema such as this one and the many others that it spawned by itself. Not only, but this artifactual model has a basis in natural forms and phenomena. It has been found that a picture works best if it contains an odd number of items in it, since it appears to be harder for the eye and the brain to fit odd numbers of items into neat groups. So, we spend more time looking at a picture with odd items in it than at one with even numbers. Thus, the Pythagorean classificatory schema turns out to have more in it than at first seemed to be there. Pythagoras also introduced the distinction between *prime* and *composite* numbers. It is a moot point to indicate that the mathematical discoveries this ancient notion has brought about, not to mention branches of mathematics, are too numerous to mention. At the basis of these primary classifications is oppositional structure which, as discussed (Chapter 3), is a *prima facie* feature of modeling systems. It would appear, at least on a superficial level, that opposition-based models such as these are part of the human brain and are projected by some unconscious force onto the world of reality. In this way, we discover things that otherwise we would not. As Umberto Eco (1998) has convincingly argued with reference to scientific discovery in general, we come upon new ideas through serendipity, which is the process of unexpectedly discovering ideas through inferences based on previous ideas and their uses. Although he does not refer to MST, it is obvious that Eco is espousing it, since serendipities would not occur without connective reasoning in the first place.

Discovery, in other words, cannot be forced. It simply happens after ideas that are given representational form subsequently suggest other ideas. By exploring the latter we come up with discoveries. Models are thus both encoders and guides of reality. St. Augustine characterized this aspect of human semiosis as a blending of our experience of signs *with* conventionalized knowledge. Discovery happens after ideas that are given formal structure subsequently suggest other ideas *through that structure*. The invention of logarithms, as discussed, came about subsequent to the invention of exponential notation, that is, the notation suggested other ways to carry out computations. These other ways led to logarithm theory which, in turn, has been an intrinsic part of mathematics ever since. Remarkably, logarithms are also found serendipitously to inhere in natural phenomena. The equation of the logarithmic spiral appears unexpectedly outside of mathematics. To quote Banks (1999: 133):

> This beautiful curve makes its appearance in many places in nature. In hydrodynamics, for example, the logarithmic spiral is produced when a vortex flow is combined with a source or sink flow – like the spiral you get when you drain the bathtub.

Models are both encoders of information and guides to the knowledge inherent in the information. It is not mere happenstance that compression and symbolization brought about the modern form of algebra. The ancient Egyptians and Babylonians used a proto-form of algebra, and hundreds of years later, so too did the Greeks, Chinese, and people of India. Diophantus used what we now call quadratic equations and symbols for unknown quantities. But between 813 and 833, al-Khwarizmi, a teacher in the mathematical school in Baghdad, wrote an influential book on algebra that came to be used as a textbook. The English word *algebra* comes from an Arabic word meaning restoration or completion in the title of this work. Restoration and completion were symbol-manipulating techniques. As such, they enshrined algebra as a separate and powerful branch of arithmetic. Today, we would see them as examples of doing the same things to both sides of an equation – an equation as a balance diagram (Chapter 3). Shortly thereafter, algebra developed into the equation modeling system that it has become. This happened between the fifteenth and seventeenth centuries when, as Bellos (2010: 123) puts it, "mathematical sentences moved from rhetorical to *symbolic* expression." As Bellos (2010: 124) goes on to write:

> Replacing words with letters and symbols was more than convenient shorthand. The symbol *x* may have started as an abbreviation for "unknown quantity", but once invented, it became a powerful tool for thought. A word or an abbreviation cannot be subjected to mathematical operation in the way that a symbol like *x* can be. Numbers made counting possible; but letter symbols took mathematics into a domain far beyond language.

Algebraic language is the language of formulas. Formulas are models that compress information. As Crilly (2011: 104) observes, the "desire to find a formula is a driving force in science and mathematics". Perhaps the world's most famous example of this is Einstein's $E = mc^2$, which compresses so much information in it that it defies common sense even to start explaining why this should be so. The formula, devised in 1905, tells us that the energy (E) into which a given amount of matter can change equals the mass (m) of that matter multiplied by the speed of light squared (c^2). Using this equation, scientists determined that the fissioning of 0.45 kilograms of uranium would release as much energy as 7,300 metric tons of TNT.

Constructing a formula is, in effect, devising a model of something. Practical experience made it obvious to the ancients that the volume of a sphere depends on the length of its radius because if one increases the radius the volume increases proportionately. Some early work showed that the increase involves multiplying the area of the base circle (obtained by multiplying the square of the radius by π (that is, πr^2). It was Archimedes who then ventured to propose that this is two-thirds of the volume of a cylinder that encloses it. And, as it turns out, he was correct.

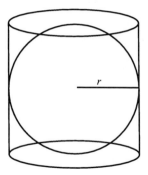

So, the height of the cylinder is the diameter, 2r, and the volume of the sphere is $2/3 \times \pi r^2 \times 2r$, which can be reduced to $4/3\pi r^3$. Now, this brilliant formula, arrived at no doubt by contemplating two diagrams at once – the sphere and the cylinder – is now part and parcel of theoretical branches, from number theory to the integral calculus.

5.3. Interconnectedness

The MST framework has allowed us to investigate how models are interconnected to produce results that would otherwise go unnoticed. There is, as discussed by Lotman and others, an implicit *interconnectedness principle* in semiotic systems which posits that expressions, symbols, representations, and traditions are interconnected to each other through connective reasoning. Such reasoning allows us to make the world understandable in concrete ways. The exploratory power of scientific models, from geometry to quantum physics, suggests that we are probably programmed to discover things through our sign structures. In observing and representing the facts of existence, thus giving them human form, we constantly stumble across hidden patterns in those facts.

Take, as an example, the case of combinatorics, the branch of mathematics that calculates possible permutations and combinations of objects. For example, three objects *a*, *b*, and *c*, can be combined (permuted) in $3 \times 2 \times 1 = 6$ ways: *abc, acb, bac, bca, cab, cba*. In this case a formula is hardly necessary. However, if we were to consider ten objects, the result would be 3,628,800; for a hundred objects it would be a mind-boggling 9.33×10^{157}. In this case a formula would increase computational economy considerably. The formula that does this is called *Stirling's formula*, named after Scottish mathematician John Stirling (1692–1770):

$$\sqrt{2\pi n} \times n^n \times e^{-n}$$

The *n* here stands for number of objects and the *e* is Euler's constant (2.718…). There are other versions of this formula, but this is one of the neatest and most economical ones around. What is e doing here, since it is associated with growth, not permutations? And what is π also doing here, associated as it is with the circle? This formula is "a reminder," writes Crilly (2011: 107), "of the amazing connections that mathematics springs on us, especially since the original problem involves only the multiplication of whole numbers." Similar examples of interconnectedness are found through the arc of mathematical history.

Interconnectedness exists both internally – among artifactual models – and externally between artifactual and natural models. One of the best known examples of the latter is the central part of Newton's gravitation theory. Newton reasoned that any two objects, with mass m_1 and m_2, are attracted to each other with a force F that is proportional to the product $m_1 \times m_2$, and divided by the square of the distance between the two masses, r^2. G is the gravitational constant, whose value depends on the chosen units of measurement:

$$F = G \frac{m_1 m_2}{r^2}$$

Perhaps no other formula has unified the behavior of natural phenomena as has this one. The interconnectedness of formula models like this one has often led to speculation if there is an overall "magic formula" for describing the universe. But, needless to say, the search for such a formula, or general theory, has been illusory. Nevertheless, some formulas appear serendipitously that seem to have "magic" properties in them. As a case in point, take a famous mathematical puzzle formulated by Leonardo Fibonacci (c. 1170-c. 1240) in his *Liber abaci*, published in 1202. As Devlin (2011: 6) has argued, the spelling should actually be *Liber abbaci*, although this is not the one commonly used in writings on Fibonacci. Fibonacci designed his book as a practical introduction to the Hindu-Arabic numeral system, which he had learned to use during his extensive travels in the Middle East. He wanted to show Europeans, who were accustomed to using the cumbersome Roman numeral system, how simple it was to solve simple problems with the Hindu-Arabic system. The puzzle in question is found in the third section of the book:

> A certain man put a pair of rabbits, male and female, in a very large cage. How many pairs of rabbits can be produced in that cage in a year if every month each pair produces a new male-female pair which, from the second month of its existence on, also is productive?

The solution goes as follows. There is 1 pair of rabbits in the cage at the start. At the end of the first month, there is still only 1 pair, for the puzzle states that a pair is productive only "from the second month of its existence on". It is during the second month that the original pair will produce its first offspring pair. Thus, at the end of the second month, a total of 2 pairs, the original one and its first offspring pair, are in the cage. Now, during the third month, only the original pair generates another new pair. The first offspring pair must wait a month before it becomes productive. So, at the end of the third month, there are 3 pairs in total in the cage – the initial pair, and the two offspring pairs that the original pair has thus far produced. If we keep tabs on the situation month by month, we can show the sequence of pairs that the cage successively contains as follows: 1, 1, 2, 3. The first digit represents the number of pairs in the cage at the start; the second, the number after one month; the third, the number after two months; and the fourth, the number after three months.

During the fourth month, the original pair produces yet another pair; so too does the first offspring pair. The other pair has not started producing yet. Therefore, during that month, a total of 2 newborn pairs of rabbits are added to the cage. Altogether, at the end of the month there are the previous 3 pairs plus the 2 newborn ones, making a total of 5 pairs in the cage. This number can now be added to our sequence: 1, 1, 2, 3, 5. During the fifth month, the original pair produces another newborn pair; the first offspring pair produces a pair as well; and now the second offspring pair produces its own first pair. The other rabbit pairs in the cage have not started producing offspring yet. So, at the end of the fifth month, 3 newborn pairs have been added to the 5 pairs that were previously in the cage, making the total number of pairs in it 5 + 3 = 8. We can now add this number to our sequence: 1, 1, 2, 3, 5, 8. Continuing to reason in this way, it can be shown that after twelve months, there are 377 pairs in the cage. Now, the intriguing thing about this puzzle is the sequence of pairs itself, if the individual computations are laid out in sequential form.

1, 1, 2, 3, 5, 8, 13, 21, 34, 55, 89, 144, 233, 377

It is interesting to note that Fibonacci himself showed the solution in the form of a layout table, which records the rabbit population growth each month:

beginning	1
first	2
second	3
third	5
fourth	8
fifth	13
sixth	21
seventh	34
eighth	55
ninth	89
tenth	144
eleventh	233
twelfth	377

Following the reasoning described above, which involves adding pairs in doublets, it becomes obvious that the salient characteristic of this sequence is that each number in it is the sum of the previous two: for example, 2 (the third number) = 1 + 1 (the sum of the previous two); 3 (the fourth number) = 1 + 2 (the sum of the previous two); etc. This pattern can of course be generalized by using simple algebraic notation. If we let F_n represent any "Fibonacci number", and F_{n-1} the number just before it, and F_{n-2} the number just before that, the pattern inherent in the sequence can be shown generally as:

$$F_n = F_{n-1} + F_{n-2}$$

This formula symbolizes the structure implicit in the sequence. It brings out the power of algebraic representation. It also shows something that may not have been apparent in the original puzzle, namely that the sequence can be extended ad infinitum, by applying the rule of continually adding the two previous numbers to generate the one after:

$$\{1, 1, 2, 3, 5, 8, 13, 21, 34, 55, 89, 144, 233, 377, 610, 987, \ldots\}$$

Fibonacci himself noted the general rule in the sequence, but little did he know how significant his sequence would become. Over the years, the properties of the Fibonacci numbers have been studied extensively, resulting in a considerable literature. Various patterns hidden in the sequence were detected and studied by the French-born mathematician Albert Girard in 1632. To get a sense at how pattern can be detected in the sequence, consider that if the nth number in the sequence is x, then every n^{th} number after x turns out to be a multiple of x:

- the third number is 2, and every third number after 2 (8, 34, 144, ...) is a multiple of 2;
- the fourth number is 3, and every fourth number after 3 (21, 144, 987, ...) is a multiple of 3;
- the fifth number is 5, and every fifth number after 5 (55, 610, ...) is a multiple of 5;
- and so on.

In the nineteenth century the term *Fibonacci sequence* was coined by the French mathematician Edouard Lucas and mathematicians began to discover myriads of patterns hidden within in it. Not only, but stretches of the sequence started cropping up in nature – in the spirals of sunflower heads, in pine cones, in the regular descent (genealogy) of the male bee, in the logarithmic (equiangular) spiral in snail shells, in the arrangement of leaf buds on a stem, in animal horns, in the botanical phenomenon known as phyllotaxis whereby the arrangement of the whorls on a pinecone or pineapple, in the petals on a sunflower, in the branches of some stems, and so on and so forth. In most flowers, for example, the number of petals is one of: 3, 5, 8, 13, 21, 34, 55, or 89 (lilies have 3 petals, buttercups 5, delphiniums 8, marigolds 13, asters 21, daisies 34 or 55 or 89). In sunflowers, the little florets that become seeds in the head of the sunflower are arranged in two sets of spirals: one winding in a clockwise direction, the other counterclockwise. The number in the clockwise is often 21, 34 and counterclockwise 34, 55, sometimes 55 and 89, and sometimes 89 and 144 in the spirals of sunflower heads (Posamentier, Lehmann 2007). As Banks (1999: 84) writes:

> The diversity of places where the Fibonacci numbers make appearances is absolutely incredible. In one form or another they show up not only in numerous topics in mathematics but also in biology, botany, music, art, and architecture.

The list of appearances is truly astounding – so much so that a journal, called *The Fibonacci Quarterly*, was founded in 1963 to publish discoveries related to, and discussions of, the sequence. Why would the solution to a simple puzzle produce numbers that are interconnected with patterns in nature and human life? There is, to the best of our knowledge, no definitive answer to this question. Maybe the appearance of formulas, such as the one describing the Fibonacci sequence, is a manifestation of the interconnectedness that exists between the human brain and reality, uncovering things about reality that would otherwise remain hidden. As mathematician Ian Stewart (2001: v) has elegantly put it, "simple puzzles could open up the hidden depths of the universe". As a play on artificial rabbit copulation the puzzle has led to an incredible discovery – namely that a simple pattern might constitute the fabric of a large slice of nature.

Perhaps Fibonacci saw something in a rabbit pen that triggered an insight into natural phenomena, leading to his puzzle, and to the hidden meanings that it contains. But that is unlikely. Fibonacci was hardly a biologist. He wanted to show how easy it is to keep track of growth through the use of decimal numbers. The Pythagoreans, as discussed a number of times, believed that mathematics mirrored the inherent order in the universe. Interestingly, the same sequence appeared first in the *Chandahshastra* (the art of prosody) written by the Indian scholar Pingala in the fourth century BCE (Devlin 2011: 145). And then, in the sixth century, the sequence was found to occur in the analysis of prosodic structure by the Indian mathematician Virahanka. Around 1159 the Jain philosopher Hemachandra wrote a work on them. But it is unlikely that Fibonacci knew about these, given the fact that knowledge of different languages at the time was minimal and scholarship did not cross borders easily.

What can be said is that the appearance of the sequence in prosody and in an artificial puzzle, as well as in nature and various human forms, constitutes a truly profound paradox. Why do mathematical formulas produce insights into external reality, even when they were not created to do so? Perhaps the ancient myths provide the only plausible response to this question. According to the *Theogony* of the Greek poet Hesiod (eighth century BCE), Chaos generated Earth, from which arose the starry, cloud-filled Heaven. In a later myth, Chaos was portrayed as the formless matter from which the Cosmos was created. In both versions, it is obvious that the ancients felt deeply that order arose out of chaos. For some truly mysterious reason, our mind requires that there be order within apparent disorder. Mathematical forms give order to the world and consequently allow us to perceive it in those (our) terms. Scientific theories and mathematical formulas are miniature blueprints of how pattern is wired into the brain and of how we search for answers to existential questions through interconnected reasoning. This is not to imply that discoveries in mathematics and science are just curious figments of mind. Rather, they are elusive bits of evidence of a theory of the world that is lurking around somewhere, but that seems to evade articulation.

Perhaps one of the most astounding patterns hidden in the sequence is the golden ratio, a fact discovered in 1753 by the Scottish mathematician Robert Simson (1687–1768), who noted that, as the numbers increased in magnitude, the ratio between succeeding numbers approached the golden ratio, whose value is .618.... The golden ratio, as we saw, results from dividing a line segment in such a way that the ratio of the whole segment to the larger part is equal to the ratio of the larger part to the smaller part. For example, if we take the stretch of numbers in the Fibonacci sequence starting with 5 and ending with 34, and take successive ratios the results are:

$$
\begin{aligned}
3/5 &= .6 \\
5/8 &= .625 \\
8/13 &= .615 \\
13/21 &= .619 \\
21/34 &= .618
\end{aligned}
$$

Why this is so is a matter of conjecture. This raises a rather profound question: are all mathematical ideas artifacts that hide features of reality in them because perhaps the human brain is itself a product of the same reality? Consider π one more time. What does it stand for? Initially, it stood simply for the fact that the ratio of the diameter to the circumference in a circle remains constant – as one increases or decreases, so does the other proportionately in the same ratio. But, then, why does π keep cropping up unexpectedly in the patterns of nature (as discussed), reminding us that it is also "out there", and defying us to understand why. Very much like the universe itself, the more sophisticated our knowledge of π grows, the more its mysteries grow. There is a beauty to π that keeps our interest in it. One can argue that π is one of those products of human effort that is a mirror of human history – it starts out in one domain of activity (geometry) and ends up in others and is probably everywhere (if we look for it). The discovery of π is a classic exemplar of the *verum-factum* principle in philosophy. Although there are precedents for it, no one was able to discuss it as insightfully as Giambattista Vico did. This principle can be defined as the ability of the human imagination to discover patterns in the world because the human mind already has such patterns built into it. The mind makes its truth, but the truth it makes then somehow makes its way into the world and is found there. As Thomas Goddard Bergin and Max Harold Fisch have perceptively pointed out, Vico believed that human beings, in being makers of things, were themselves made to do just that: "Men have themselves made this world of nations, but it was not without drafting, it was even without seeing the plan that they did just what the plan called for." (Vico 1984: xlv). As Peirce similarly put it, the mind has "a natural bent in accordance with nature" (CP 6.478). This blending of mind and nature becomes perception, which Peirce called the "outward clash" of the physical world on the senses. All this is not as far-fetched as it might seem, since today quantum physics is basically a science that puts the verum-factum principle to its most interesting test. Perhaps we do influence the world with our minds, even giving it the structure that we believe we are describing; on the other hand, maybe that is exactly how things are.

In observing the facts of existence, we constantly stumble across hidden patterns. A similar argument could be made for a whole host of metaforms, such as e, $e^{i\pi} + 1 = 0$, among many others, which have turned out to have a wide variety of serendipitous applications. As mentioned, the number e was discovered by Euler in 1727 as the limit of the expression $(1 + 1/n)^n$ as n becomes large without bound. Its limiting value is approximately 2.7182818285. Unlike π, e has no simple geometric interpretation. Yet it forms the base of natural logarithms; it appears in the fundamental function for equations describing growth and many other processes of change; it surfaces as well in the formulas for many curves; it crops up frequently in the theory of probability and in formulas for calculating compound interest; and the list could go on. Now, why Euler devised that formula in the first place is not clear. He certainly could not have known the kinds of ideas and applications it would have led to, since these came after its formulation.

5.4. Mathematics and reality

The word *serendipity* was coined by Horace Walpole in 1754, from the title of the fairy tale *The Three Princes of Serendip*, whose heroes make many fortunate discoveries accidentally (Merton, Barber 2003). As it turns out, they made the discoveries on the basis of what Walpole called "accidental sagacity", which tallies with the Peircean concept of abduction and the Vichian one of *ingegno* (ingenuity based on imagination). This type of sagacity consists in inferring events and patterns on the basis of previous experience. Ceylon's ancient name was Serendip, and it was Walpole who, after having read the tale, introduced the word *serendipity* into the English language. This term characterizes the history of discovery in mathematics and science – for example, Röntgen accidentally discovered X-rays by seeing their effects on photographic plates; Fleming stumbled upon penicillin by noticing the effects of a mold on bacterial cultures; and so on (Roberts 1989). Perhaps the most famous of all serendipitous episodes in the history of science is Archimedes' discovery of a law of hydrostatics as he was supposedly taking a bath. After grasping the law in his mind through a flash of insight, he is purported to have run out into the streets of Syracuse naked, crying "Eureka", meaning "I have found it". Since then, such flashes of insight have been called "Eureka moments".

The tale goes somewhat as follows. Three princes from Ceylon were journeying in a strange land when they came upon a man looking for his lost camel. The princes had never seen the animal, but they asked the owner a series of seemingly pertinent questions: Was it missing a tooth? Was it blind in one eye? Was it lame? Was it laden with butter on one side and honey on the other? Was it being ridden by a pregnant woman? Incredibly, the answer to all their questions was yes. The owner instantly accused the princes of having stolen the animal since, clearly, they could not have had such precise knowledge otherwise. But the princes merely pointed out that they had observed the road, noticing that the grass on either side was uneven and this was most likely the result of the camel eating the grass. They had also noticed parts of the grass that were chewed unevenly, suggesting a gap in the animal's mouth. The uneven patterns of footprints indicated signs of awkward mounting and dismounting, which could be related to uneven weights on the camel. Given the society of the era, this suggested the possibility that the camel was ridden by a pregnant woman, creating a lack of equilibrium and thus an uneven pattern of footprints. Finally, in noticing differing accumulations of ants and flies they concluded that the camel was laden with butter and honey – the natural attractors of these insects. Their questions were, as it turns out, inferences based on astute observations. So, serendipity is not a mysterious phenomenon after all; it comes from connecting experiences and mental forms. Serendipity is the result of abduction, to use Peircean theory:

> The abductive suggestion comes to us like a flash. It is an act of insight, although of extremely fallible insight. It is true that the different elements of the hypothesis were in our minds before; but it is the idea of putting together what we had never before dreamed of putting together which flashes the new suggestion before our contemplation. (CP 5.180)

As discussed throughout this book, serendipity plays a role in the manifestations of a theoretical form in domains other than the original one in which it was forged. As just mentioned, π appears in a number of mathematical calculations and formulas, such as the one that describes the motion of a pendulum or the vibration of a string (Beckmann 1971; Blatner 1997; Eymard, Lafon 2004; Posamentier 2004). It also turns up in equations describing the DNA double helix, rainbows, ripples spreading from where a raindrop falls into water, all kinds of waves, navigation systems, and on and on. All that can be said is that π is both invented and discovered (Menninger 1969). The human mind invents or creates numbers and other mathematical constructs in the same sense that it creates colors and other sign forms. Why these often lead to discoveries not intended in the initial act of construction is still a mystery. And they spur us on to look for a magic formula to everything, even if there probably isn't one.

Mathematical models allow us to represent the world in various ways. At the same time, they serendipitously unravel patterns within nature itself. This synergy between the semiosphere and the biosphere is a remarkable one indeed. From the dawn of civilization to the present age, it has always been felt that there is an intrinsic connection between the two. The *raison d'être* of semiotics is, arguably, to investigate whether or not reality can exist independently of the signs that human beings create to represent and think about it. Is the physical universe a great machine operating according to natural laws that may be discovered by human reason? Or, on the other hand, is everything out there no more than a construction of the human mind deriving its categories from the world of sensations and perceptions? Although an answer to this fundamental question will clearly never be possible, one of the important offshoots of the search for an answer has been a systematic form of inquiry into how the mind's products and the body's natural processes are interrelated.

5.5. Concluding remarks

In his monumental history of semiotics, John Deely (2001) argues essentially that we can only understand the history of knowledge by mapping it against the development of sign theory. Our excursion into the nature of mathematics as a modeling system bears this out perfectly. Perhaps, as Crilly (2011: 153) observes, mathematical models are unwitting mirrors of things in the world:

> A mathematical model is a way of describing a real-life situation in mathematical language, turning it into the vocabulary of variables and equations. By making assumptions about what is important and ignoring some particulars that may be disregarded, the model aims to capture *essentials* of a

situation. A mathematical mechanism is set up that can then be validated, in order to check whether the model mirrors the real-life situation.

The time has come to unite mathematics and semiotics. The unification will make questions such as the ones raised in this book more tangible and open to inquiry by scholars in both fields working together. From the Pythagorean practice of giving sacrifice to the gods for mathematical discoveries to the seventeenth century practice on the part of the Japanese of giving *sangaku* (the Japanese word for "mathematical tablet") to the spirits for discovering mathematical proofs, there seems to be a universal feeling across the world that discoveries reveal the world to us in bits and pieces. This is why the ancients thought that a causal connection existed between earthly matters and the stars. Those who could use numbers to calculate forthcoming events, such as the next planting season, garnered great power unto themselves, becoming wizards, mathematicians, and astronomers. MST provides a framework for understanding why mathematics works.

But even MST really does not penetrate the substance of the enigma at hand. Nor does it really answer the question of the interconnection between mathematical models and their serendipitous appearance in the world. Semiotics is a descriptive science, after all, not an explanatory one. So we are left with the same kinds of questions with which we started off this book: Why does mathematics work as a model to explain the physical world? Why is the Pythagorean theorem, for instance, real, explaining a whole range of phenomena? As pointed out in the first chapter, the historian of science Jacob Bronowski characterized the as the most important single theorem in the whole of mathematics, because it sets the stage for using theorems for modeling the world in which we live. If space had a different symmetry the theorem would not be true. As Clawson (1999: 284) has suggested, mathematics might even explain the laws of unknown universes: "Certain mathematical truths are the same beyond this particular universe and work for all potential universes". This book has only skimmed the surface of the implications that studying mathematics from the perspective of MST holds for all of us.

As mentioned a number of times, when Gödel showed rather matter-of-factly that there never can be a consistent system of axioms that can capture all the truths of mathematics, he showed, in effect, that the *makers* of the axioms could never extricate themselves from *making* them. Gödel made it obvious to mathematicians that mathematics was made by them, and that the exploration of "mathematical truth" would go on forever as long as humans were around. The final map of the mathematical realm will never be drawn. Like other products of the imagination, the world of mathematics lies within the minds of humans. In effect, all sign systems are "theories" of reality, evaluating it in their own particular ways. In other words, our knowledge systems can only give us partial glimpses of reality.

References

Aczel, Amir D. 2000. *The Mystery of the Aleph: Mathematics, the Kabbalah and the Search for Infinity*. New York: Four Walls Eight Windows.

Adam, John A. 2004. *Mathematics in Nature: Modeling Patterns in the Natural World*. Princeton: Princeton University Press.

Alexander, James C. 2012. On the cognitive and semiotic structure of mathematics. In: Bockarova, Mariana; Danesi, Marcel; Núñez, Rafael (eds.), *Semiotic and Cognitive Science Essays on the Nature of Mathematics*. Munich: Lincom Europa, 1–34.

Alpher, Barry 1987. Feminine as the unmarked grammatical gender: Buffalo girls are no fools. *Australian Journal of Linguistics* 7: 169–187.

Andersen, Henning 1989. Markedness theory: The first 150 years. In: Tomic, Olga Miseska (ed.), *Markedness in Synchrony and Diachrony*. Berlin: Mouton de Gruyter, 11–16.

– 2001. Markedness and the theory of linguistic change. In: Andersen, Henning (ed.), *Actualization. Linguistic Change in Progress*. Amsterdam: John Benjamins, 19–57.

– 2008. Naturalness and markedness. In: Willems, Klaas; De Cuypere, Ludovic (eds.), *Naturalness and Iconicity in Language*. Amsterdam: John Benjamins, 101–119.

Anderson, Myrdene; Sáenz-Ludlow, Adalira; Cifarelli, Victor V. 2000. Musement in mathematical manipulation. In: Gimate-Welsh, Adrián S. (ed.), *Ensayos Semióticos*. Mexico: Porrúa, 663–676.

– (eds.) 2003. *Educational Perspectives on Mathematics as Semiosis: From Thinking to Interpreting to Knowing*. Ottawa: Legas Press.

Andrews, Edna 1990. *Markedness Theory: The Union of Asymmetry and Semiosis in Language*. Durham: Duke University Press.

– 2003. *Conversations with Lotman: Cultural Semiotics in Language, Literature, and Cognition*. Toronto: University of Toronto Press.

Andrews, Edna; Tobin, Yishai (eds.) 1996. *Toward a Calculus of Meaning: Studies in Markedness, Distinctive Features and Deixis*. Amsterdam: John Benjamins.

Anfindsen, Jens Tomas 2006. *Aristotle on Contrariety as a Principle of First Philosophy*. Uppsala: Department of Philosophy, Uppsala University.

Arranz, José I. Prieto 2005. Towards a global view of the transfer phenomenon. *The Reading Matrix* 5: 116–128.

Asch, Solomon E. 1950. On the use of metaphor in the description of persons. In: Werner, Heinz (ed.), *On Expressive Language*. Worcester: Clark University Press, 86–94.

– 1958. The metaphor: A psychological inquiry. In: Tagiuri, Renato; Petrullo, Luigi (eds.), *Person Perception and Interpersonal Behavior*. Stanford: Stanford University Press, 28–42.

Askew, Mike; Ebbutt, Sheila 2011. *The Bedside Book of Geometry: From Pythagoras to the Space Race: The ABC of Geometry*. London: New Burlington Books.

Babin, Arthur Eugène 1940. *The Theory of Opposition in Aristotle*. Notre Dame: University of Notre Dame.

Ball, Keith 2003. *Strange Curves, Counting Rabbits, and Other Mathematical Explorations*. Princeton: Princeton University Press.

Ball, Walter William Rouse 1892. *Mathematical Recreations and Problems of Past and Present Times*. London: Macmillan.

Banks, Robert B. 1999. *Slicing Pizzas, Racing Turtles, and Further Adventures in Applied Mathematics*. Princeton: Princeton University Press.

Barbaresi, Lavinia Merlini 1988. *Markedness in English Discourse: A Semiotic Approach*. Parma: Edizioni Zara.

Barker-Plummer, Dave; Bailin, Sidney C. 1997. The role of diagrams in mathematical proofs. *Machine Graphics and Vision* 8: 25–58.

– 2001. On the practical semantics of mathematical diagrams. In: Anderson, Michael (ed.), *Reasoning with Diagrammatic Representations*. New York: Springer.

Barnsley, Michael F. 1988. *Fractals Everywhere*. Boston: Academic.

Beckmann, Petr 1971. *A History of π*. New York: St. Martin's Press.

Belardi, Walter 1970. *L'opposizione privativa*. Napoli: Istituto Universitario Orientale di Napoli.

Bellos, Alex 2010. *Here's Looking at Euclid: A Surprising Excursion through the Astonishing World of Math*. Princeton: Princeton University Press.

Belsey, Catherine 2002. *Poststructuralism: A Very Short Introduction*. Oxford: Oxford University Press.

Benveniste, Émile 1946. Structure des relations de personne dans le verbe. *Bulletin de la Société de Linguistique de Paris* 43: 225–236.

Beziau, Jean-Yves; Payette, Gillman (eds.) 2012. *The Square of Opposition: A General Framework for Cognition*. New York: Peter Lang.

Bing, Thomas J.; Redish, Edward F. 2007. The cognitive blending of mathematics and physics knowledge. *2006 Physics Education Research Conference Proceedings* 883: 26–29.

Black, Max 1962. *Models and Metaphors. Studies in Language and Philosophy*. Ithaca: Cornell University Press.

Blanché, Robert 1966. *Structures intellectuelles: Essai sur l'organisation systématique des concepts*. Paris: Vrin.

Blatner, David 1997. *The Joy of Pi*. Harmondsworth: Penguin.

Bocheński, Józef Maria 1961. *A History of Formal Logic*. [Thomas, Ivo, trans.] Notre Dame: University of Notre Dame Press.

Bockarova, Mariana; Danesi, Marcel; Núñez, Rafael (eds.) 2012. *Semiotic and Cognitive Science Essays on the Nature of Mathematics*. Munich: Lincom Europa.

Bogoslovksy, Boris Basil 1928. *The Technique of Controversy: Principles of Dynamic Logic*. London: Paul, Trench and Teubner.

Bolinger, Dwight 1968. *Aspects of Language*. New York: Harcourt, Brace, Jovanovich.

Bronowski, Jacob 1973. *The Ascent of Man*. Boston: Little, Brown, and Co.

Brown, Roger 1958a. How shall a thing be called? *Psychological Review* 65: 14–21.

– 1958b. *Words and Things: An Introduction to Language*. New York: The Free Press.

Brown, Roger W.; Leiter, Raymond A.; Hildum, Donald C. 1957. Metaphors from music criticism. *Journal of Abnormal and Social Psychology* 54: 347–352.

Bühler, Karl 1934. *Sprachtheorie: Die Darstellungsfunktion der Sprache*. Jena: Fischer.

– 1951[1908]. On thought connection. In: Rapaport, David (ed.), *Organization and Pathology of Thought*, 81–92. New York: Columbia University Press.

Bunt, Lucas N. H.; Jones, Phillip S.; Bedient, Jack D. 1976. *The Historical Roots of Elementary Mathematics*. New York: Dover.

Cherry, Colin 1966. *On Human Communication*. Cambridge: MIT Press.

Cho, Yang Seok; Proctor, Robert W. 2007. When is an odd number not odd? Influence of task rule on the MARC effect for numeric classification. *Journal of Experimental Psychology. Learning, Memory, and Cognition* 33: 832–842.

Chomsky, Noam 1957. *Syntactic Structures*. The Hague: Mouton.

Chomsky, Noam; Halle, Morris 1968. *The Sound Pattern of English*. New York: Harper & Row.

Clawson, Calvin C. 1999. *Mathematical Sorcery: Revealing the Secrets of Numbers*. Cambridge: Perseus.

Cole, K. C. 1984. *Sympathetic Vibrations: Reflections on Physics as a Way of Life*. New York: Bantam.

Collins, Joan M. 1969. *An Exploration of the Role of Opposition in Cognitive Processes of Kindergarten Children*. Toronto: Ontario Institute for Studies in Education Theory.

Coseriu, Eugenio 1973. *Probleme der strukturellen Semantik*. Tübingen: Tübinger Beiträge zur Linguistik 40.

CP = Peirce, Charles 1931–1958.

Crilly, Tony 2011. *Mathematics*. London: Quercus.

Danesi, Marcel 1993. *Vico, Metaphor, and the Origin of Language*. Bloomington: Indiana University Press.

– 2002. *The Puzzle Instinct: The Meaning of Puzzles in Human Life*. Bloomington: Indiana University Press.

– 2004a. *Poetic Logic: The Role of Metaphor in Thought, Language, and Culture*. Madison: Atwood Publishing.

– 2004b. *Messages, Signs, and Meanings: A Basic Textbook in Semiotics and Communication Theory* [3rd ed.]. Toronto: Canadian Scholars' Press,

– 2007. *The Quest for Meaning: A Guide to Semiotic Theory and Practice*. Toronto: University of Toronto Press.

– 2008. *Problem-Solving in Mathematics: A Semiotic Perspective for Educators and Teachers*. New York: Peter Lang.

Danesi, Marcel; Maida-Nicol, Sara (eds.) 2012. *Foundational Texts in Linguistic Anthropology*. Toronto: Canadian Scholars Press.

Davis, Phillip J.; Hersh, Reuben 1986. *Descartes' Dream: The World According to Mathematics*. Boston: Houghton Mifflin.

Deely, John 2001. *Four Ages of Understanding: The First Postmodern Survey of Philosophy from Ancient Times to the Turn of the Twentieth Century*. Toronto: University of Toronto Press.

Dehaene, Stanislas 1997. *The Number Sense: How the Mind Creates Mathematics*. Oxford: Oxford University Press.

Derrida, Jacques 1967. *De la grammatologie*. Paris: Minuit.

Devlin, Keith 1997. *Mathematics: The Science of Patterns*. New York: Scientific American Library.

– 2011. *The Man of Numbers: Fibonacci's Arithmetic Revolution*. New York: Walker and Company.

Dirven, René; Verspoor, Marjolijn (eds.) 2004. *Cognitive Exploration of Language and Linguistics*. Amsterdam: John Benjamins.

Dunlap, Richard A. 1997. *The Golden Ratio and Fibonacci Numbers*. Singapore: World Scientific.

Eckman, Fred R.; Moravcsik, Edith A.; Wirth, Jessica R. (eds.) 1983. *Markedness*. New York: Plenum.

Eco, Umberto 1998. *Serendipities: Language and Lunacy* [Weaver, William, trans.]. New York: Columbia University Press.

Elšík, Viktor; Matras, Yaron 2006. *Markedness and Language Change: The Romani Sample*. Berlin: Mouton de Gruyter.

Emmeche, Claus; Kull, Kalevi (eds.) 2011. *Towards a Semiotic Biology: Life is the Action of Signs*. London: World Scientific.

English, Lyn D. (ed.) 1997. *Mathematical Reasoning: Analogies, Metaphors, and Images*. Mahwah: Lawrence Erlbaum Associates.

Eymard, Pierre; Lafon, Jean-Paul 2004. *The Number Pi*. [Wilson, Stephen S., trans.] New York: American Mathematical Society.

Fauconnier, Gilles; Turner, Mark 2002. *The Way We Think: Conceptual Blending and the Mind's Hidden Complexities*. New York: Basic.

Ferrer i Cancho, Ramon 2005. The variation of Zipf's law in human language. *European Physical Journal* 44: 249–257.

Ferrer i Cancho, Ramon; Solé, Ricard V. 2001. Two regimes in the frequency of words and the origins of complex lexicons: Zipf's law revisited. *Journal of Quantitative Linguistics* 8:165–231.

Ferrer i Cancho, Ramon; Riordan, Oliver; Bollobás, Béla 2005. The consequences of Zipf's law for syntax and symbolic reference. *Proceedings of the Royal Society of London, Series B, Biological Sciences*. London: Royal Society of London, 1–5.

Flood, Robin; Wilson, Raymond 2011. *The Great Mathematicians: Unravelling the Mysteries of the Universe*. London: Arcturus.

Foucault, Michel 1972. *The Archeology of Knowledge* [Smith, Sheridan; Mark, Alan, trans.]. New York: Pantheon.

Fox, James J. 1974. Our ancestors spoke in pairs: Rotinese views of language, dialect and code. In: Bauman, Richard; Scherzer, Joel (eds.), *Explorations in the Ethnography of Speaking*. Cambridge: Cambridge University Press, 65–88.

–   1975. On binary categories and primary symbols. In: Willis, Roy G. (ed.), *The Interpretation of Symbolism*. London: Malaby, 99–132.

Geeraerts, Dirk (ed.) 2006. *Cognitive Linguistics: Basic Readings*. Berlin: Mouton de Gruyter.

Ghyka, Matila 1977. *The Geometry of Art and Life*. New York: Dover.

Gibbs Jr., Raymond W. 1994. *The Poetics of Mind: Figurative Thought, Language, and Understanding*. Cambridge: Cambridge University Press.

Gibbs Jr., Raymond W.; Colston, Herbert L. 2012. *Interpreting Figurative Meaning*. Cambridge: Cambridge University Press.

Goatley, Andrew 1997. *The Language of Metaphors*. London: Routledge.

Gödel, Kurt 1931. Über formal unentscheidbare Sätze der Principia Mathematica und verwandter Systeme, Teil I. *Monatshefte für Mathematik und Physik* 38: 173–189.

Godel, Robert 1957. *Les sources manuscrites du "Cours de linguistique générale" de F. de Saussure*. Paris: Minard.

Godino, Juan D.; Batanero, Carmen; Font, Vicenç 2007. The onto-semiotic approach to research in mathematics education. *Mathematics Education* 39: 127–135.

Greimas, Algirdas Julius 1966. *Sémantique structurale*. Paris: Larousse.

–   1970. *Du sens*. Paris: Seuil.

–   1987. *On Meaning: Selected Essays in Semiotic Theory*. [Perron, Paul; Collins, Frank, trans.] Minneapolis: University of Minnesota Press.

Hallyn, Fernand 1990. *The Poetic Structure of the World: Copernicus and Kepler*. New York: Zone Books.

Harel, Guershon; Sowder, Larry 2007. Toward comprehensive perspectives on the learning and teaching of proof. In: Lester, Frank K. (ed.), *Second Handbook of Research on Mathematics Teaching and Learning*. Charlotte: Information Age Publishing, 805–842.

Haspelmath, Martin 2006. Against markedness (and what to replace it with). *Journal of Linguistics* 42: 25–70.

Hatten, Robert S. 2004. *Musical Meaning in Beethoven: Markedness, Correlation and Interpretation*. Bloomington: Indiana University Press.

Hertz, Robert 1973. The pre-eminence of the right hand: A study in religious polarity. In: Needham, Rodney (ed.). *Right and Left: Essays on Dual Symbolic Classification*. Chicago: University of Chicago Press, 23–36.

Hjelmslev, Louis 1939. Note sur les oppositions supprimables. *Travaux de Cercle Linguistique de Prague* 8: 51–57.

–   1959. *Essais linguistique*. Copenhagen: Munksgaard.

Hobbes, Thomas 1839[1656–1658]. *Elements of Philosophy*. London: Molesworth.

Honeck, Richard P.; Hoffman, Robert R. (eds.) 1980. *Cognition and Figurative Language*. Hillsdale: Lawrence Erlbaum Associates.

Hume, David 1902[1749]. *An Enquiry Concerning Human Understanding*. Oxford: Clarendon.

Ivanov, Viacheslav V. 1974. On antisymmetrical and asymmetrical relations in natural languages and other semiotic systems. *Linguistics* 119: 35–40.

Jakobson, Roman 1932. Zur Struktur des russischen Verbum. In: *Charisteria Guilelma Mathesio Quinquagenario a Discipulis et Circuli Linguistici Pragensis Sodalibus Oblata*. Prague: Prazsky lingvistick, 74–84.

–   1936. Beitrag zur allgemeinen Kasuslehre: Gesamtbedeutungen der russischen Kasus. *Travaux du Cercle Linguistique de Prague* 6: 244–88.

–   1939. Observations sur le classement phonologique des consonnes. *Proceedings of the Fourth International Congress of Phonetic Sciences*, Helsinki, 34–41.

–   1942. *Kindersprache, Aphasie und algemeine Lautgesetze*. Uppsala: Almqvist and Wiksell.

–   1968. The role of phonic elements in speech perception. *Zeitschrift für Phonetik, Sprachwissenschaft und Kommunikationsforschung* 21: 9–20.

Jakobson, Roman; Waugh, Linda R. 1979. *Six Lectures on Sound and Meaning*. Cambridge: MIT Press.

Jakobson, Roman; Fant, C. Gunnar M.; Halle, Morris 1952. *Preliminaries to Speech Analysis*. Cambridge: MIT Press.

Jakobson, Roman; Karcevskij, Serge; Trubetzkoy, Nikolai 1928. Proposition au premier congrès international des linguistes: Quelles sont les méthodes les mieux appropriées à un exposé complet et pratique de la phonologie d'une langue quelconque? *Premier Congrès International des Linguistes, Propositions*, 36–39.

Johnson, Mark 1987. *The Body in the Mind: The Bodily Basis of Meaning, Imagination and Reason*. Chicago: University of Chicago Press.

Jones, Roger S. 1982. *Physics as Metaphor*. New York: New American Library.

Kant, Immanuel. 1965[1781]. *Critique of Pure Reason*. [Kemp Smith, Norman, trans.]. New York: St. Martin's.

King, Ruth Elizabeth 1991. *Talking Gender: A Guide to Nonsexist Communication*. Toronto: Copp Clark Pitman Ltd.

Kline, Morris 1969. *Mathematics and the Physical World*. New York: Dover.

Koen, Frank 1965. An intra-verbal explication of the nature of metaphor. *Journal of Verbal Learning and Verbal Behavior* 4: 129–133.

Kosslyn, Stephen M. 1983. *Ghosts in the Mind's Machine: Creating and Using Images in the Brain*. New York: W. W. Norton.

– 1994. *Image and Brain*. Cambridge: MIT Press.

Kull, Kalevi (ed.) 2001. *Jakob von Uexküll: A Paradigm for Biology and Semiotics* [Special Issue]. *Semiotica* 134.

– 2012. Semiosis includes incompatibility: On the relation between semiotics and mathematics. In: Bockarova, Mariana; Danesi, Marcel; Núñez, Rafael (eds.), *Semiotic and Cognitive Science Essays on the Nature of Mathematics*. Munich: Lincom Europa, 302–334.

Kulpa, Zenon 2004. On diagrammatic representation of mathematical knowledge. In: Asperti, Andrea; Bancerek, Grzegorz; Trybulec, Andrej (eds.), *Mathematical Knowledge Management*. New York: Springer, 191–204.

Lakoff, George 1987. *Women, Fire, and Dangerous Things: What Categories Reveal about the Mind*. Chicago: University of Chicago Press.

– 2008. The Neural Theory of Metaphor. In: Gibbs, Raymond W. (ed.), *The Cambridge Handbook of Metaphor and Thought*. Cambridge: Cambridge University Press, 17–38.

– 2011. The cognitive and neural foundation of mathematics: The Case of Gödel's metaphors. Keyfitz Lecture, March 14. Toronto: Fields Institute of Mathematics, University of Toronto.

– 2012. The contemporary theory of metaphor. In: Danesi, Marcel; Maida-Nicol, Sara (eds.), *Foundational Texts in Linguistic Anthropology*. Toronto: Canadian Scholars' Press, 157–201.

Lakoff, George; Johnson, Mark 1980. *Metaphors We Live By*. Chicago: Chicago University Press.

– 1999. *Philosophy in the Flesh: The Embodied Mind and Its Challenge to Western Thought*. New York: Basic.

Lakoff, George; Núñez, Rafael E. (2000). *Where Mathematics Comes From: How the Embodied Mind Brings Mathematics into Being*. New York: Basic Books.

Langacker, Ronald Wayne 1987. *Foundations of Cognitive Grammar*. Stanford: Stanford University Press.

– 1990. *Concept, Image, and Symbol: The Cognitive Basis of Grammar*. Berlin: Mouton de Gruyter.

– 1999. *Grammar and Conceptualization*. Berlin: Mouton de Gruyter.

Langer, Susanne K. 1948. *Philosophy in a New Key*. New York: Mentor Books.

Lave, Jean 1988. *Cognition in Practice: Mind, Mathematics and Culture in Everyday Life*. Cambridge: Cambridge University Press.

Lepik, Peet 2008. *Universals in the Context of Juri Lotman's Semiotics*. Tartu: Tartu University Press.

Lesh, Richard; Harel, Guershon 2003. Problem solving, modeling, and local conceptual development. *Mathematical Thinking and Learning* 5: 157.

Lévi-Strauss, Claude 1958. *Anthropologie structurale*. Paris: Pion.

– 1971. *L'Homme nu*. Paris: Pion.

Liebeck, Pamela 1984. *How Children Learn Mathematics: A Guide for Parents and Teachers*. Harmondsworth: Penguin.

Livio, Mario 2002. *The Golden Ratio: The Story of Phi, the World's Most Astonishing Number*. New York: Broadway Books.

Locke, John 1975[1690]. *An Essay Concerning Human Understanding* [Nidditch, Peter H., ed.] Oxford: Clarendon Press.

Lorrain, François 1975. *Réseaux sociaux et classifications sociales*. Paris: Hermann.

Lotman, Yuri 1991. *Universe of the Mind: A Semiotic Theory of Culture*. Bloomington: Indiana University Press.

Lucy, John A. 1992. *Language Diversity and Thought: A Reformulation of the Linguistic Relativity Hypothesis*. Cambridge: Cambridge University Press.

MacNamara, Olwen 1996. Mathematics and the Sign. *Proceedings of PME* 20: 369–378.

Malmberg, Bertil 1976. Langue–forme–valeur: Reflexion sur trois concepts saussuriennes. *Semiotica* 18: 3.

Mandelbrot, Benoit 1977. *The Fractal Geometry of Nature*. San Francisco: Freeman.

Mansouri, Fethi 2000. *Grammatical Markedness and Information Processing in the Acquisition of Arabic [as] a Second Language*. München: Lincom Europa.

Maor, Eli 2007. *The Pythagorean Theorem: A 4,000-Year History*. Princeton: Princeton University Press.

Marcus, Solomon 1975. The metaphors and the metonymies of scientific (especially mathematical) language. *Revue Roumaine de Linguistique* 20: 535–537.

– 1980. The paradoxical structure of mathematical language. *Revue Roumaine de Linguistique* 25: 359–366.

– 2003. Mathematics through the glasses of Hjelmslev's semiotics. *Semiotica* 145: 235–246.

– 2010. Mathematics as semiotics. In: Sebeok, Thomas A.; Danesi, Marcel (eds.), *Encyclopedic Dictionary of Semiotics* [3rd edition]. Berlin: Mouton de Gruyter.

– 2012. Mathematics between semiosis and cognition. In: Bockarova, Mariana; Danesi, Marcel; Núñez, Rafael (eds.). *Semiotic and Cognitive Science Essays on the Nature of Mathematics*. Munich: Lincom Europa, 98–182.

Martinet, André 1960. *Éléments de linguistique générale*. Paris: Colin.

Mazur, Joseph 2008. *Zeno's Paradox: Unravelling the Ancient Mystery behind Space and Time*. New York: Plume.

McCarthy, John J. 2001. *A Thematic Guide to Optimality Theory*. Cambridge: Cambridge University Press.

McCowan, Brenda; Hanser, Sean F.; Doyle, Laurance R. 1999. Quantitative tools for comparing animal communication systems: Information theory applied to bottlenose dolphin whistle repertoires. *Animal Behaviour* 62: 1151–1162.

Mel'čuk, Igor 2001. *Linguistic Theory: Communicative Organization in Natural Language*. Amsterdam: John Benjamins.

Menninger, Karl 1969. *Number Words and Number Symbols: A Cultural History of Number*. Cambridge: MIT Press.

Merton, Robert K.; Barber, Elinor 2003. *The Travels and Adventures of Serendipity: A Study in Sociological Semantics and the Sociology of Science*. Princeton: Princeton University Press.

Mettinger, Arthur 1994. *Aspects of Semantic Opposition in English*. Oxford: Oxford University Press.

Miller, George Armitage 1951. *Language and Communication*. New York: McGraw-Hill.

– 1956. The magical number seven, plus or minus two: Some limits on our capacity for processing information. *Psychological Review* 63: 81–97.

– 1981. *Language and Speech*. New York: W. H. Freeman.

Mitchell, William John Thomas; Davidson, Arnold I. (eds.) 2007. *The Late Derrida*. Chicago: University of Chicago Press.

Morris, Charles 1938. *Foundations of the Theory of Signs*. Chicago: University of Chicago Press.

Müller, Cornelia 2008. *Metaphors Dead and Alive, Sleeping and Waking: A Dynamic View*. Chicago: University of Chicago Press.

Musser, Gary L.; Burger, William F.; Peterson, Blake E. 2006. *Mathematics for Elementary Teachers: A Contemporary Approach*. Hoboken: John Wiley.

Needham, Rodney 1973. *Right and Left*. Chicago: University of Chicago Press.

Neugebauer, Otto Eduard; Sachs, Abraham Joseph 1945. *Mathematical Cuneiform Texts*. New Haven: American Oriental Society.

Niederman, Derrick 2012. *The Puzzler's Dilemma: From the Lighthouse of Alexandria to Monty Hall, a Fresh Look at Classic Conundrums of Logic, Mathematics and Life*. New York: Perigree.

Nietzsche, Friedrich 1979[1873]. *Philosophy and Truth: Selections from Nietzsche's Notebooks of the Early 1870's*. Atlantic Heights: Humanities Press.

Núñez, Rafael E.; Edwards, Laurie D.; Filipe Matos, João 1999. Embodied cognition as grounding for situatedness and context in mathematics education. *Educational Studies in Mathematics* 39: 45–65.

Ogden, Charles Kay 1932. *Opposition: A Linguistic and Psychological Analysis*. London: Paul, Trench, and Trubner.

Ogden, Charles Kay; Richards, Ivor Armstrong 1923. *The Meaning of Meaning*. London: Routledge and Kegan Paul.

Ogilvy, C. Stanley 1956. *Excursions in Mathematics*. New York: Dover.

Ortony, Andrew (ed.) 1979. *Metaphor and Thought*. Cambridge: Cambridge University Press.

Osgood, Charles E.; Suci, George J. 1953. Factor analysis of meaning. *Journal of Experimental Psychology* 49: 325–328.

Osgood, Charles E.; Suci, George J.; Tannenbaum, Percy H. 1957. *The Measurement of Meaning*. Urbana: University of Illinois Press.

Otte, Michael 1997. Mathematics, semiotics, and the growth of social knowledge. *For the Learning of Mathematics* 17: 47–54.

Pappas, Theoni 1991. *More Joy of Mathematics*. San Carlos: Wide World Publishing.

– 1999. *Mathematical Footprints: Discovering Mathematical Impressions All Around Us*. San Carlos: World Wide Publishing.

Park, Hye-Sook 2000. Markedness and learning principles in SLA: Centering on acquisition of relative clauses. *Journal of Pan-Pacific Association of Applied Linguistics* 4: 87–114.

Parsons, Talcott; Bales, Robert F. 1955. *Family, Socialization, and Interaction Process*. Glencoe: Free Press.

Pavlov, Ivan 1902. *The Work of Digestive Glands*. London: Griffin.

Peirce, Charles Sanders 1931–1958. *Collected Papers of Charles Sanders Peirce*. Cambridge: Harvard University Press. Vols. 1–8. [Hartshorne, Charles; Weiss, Paul (eds.), 1931–1935; vols. 7–8. Burks, Arthur W. (ed.), 1958. In-text references are to CP, followed by volume and paragraph numbers.]

Piaget, Jean 1952. *The Child's Conception of Number*. London: Routledge and Kegan Paul.

Pimm, David 1995. *Symbols and Meanings in School Mathematics*. London: Routledge.

Pollio, Howard R; Barlow, Jack M.; Fine, Harold J.; Pollio, Marilyn R. 1977. *Psychology and the Poetics of Growth: Figurative Language in Psychology, Psychotherapy, and Education*. Hillsdale: Lawrence Erlbaum Associates.

Pos, Hendrik J. 1938. La notion d'opposition en linguistique. *Xle Congrès International de Psychologie*. Paris: Alcan, 246–247.

– 1964. Perspectives du structuralisme. In: *Études phonologiques dediées à la mémoire de M. le Prince K. S. Trubetzkoy*. [*Travaux du Cercle linguistique de Prague*, 8]. Prague: Jednota Českych Mathematiku Fysiku, 71–78.

Posamentier, Alfred S. 2004. *Pi: A Biography of the World's Most Mysterious Number*. New York: Prometheus.

Posamentier, Alfred S.; Hauptman, Herbert A. 2010. *The Pythagorean Theorem: The Story of Its Power and Beauty*. New York: Prometheus.

Posamentier, Alfred S.; Lehmann, Ingmar 2007. *The (Fabulous) Fibonacci Numbers*. New York: Prometheus.

Post, Thomas R.; Wachsmuth, Ipke; Lesh, Richard; Behr, Merlyn J. 1985. Order and equivalence of rational numbers: A cognitive analysis. *Journal for Research in Mathematics Education* 16: 18–36.

Pottier, Bernard 1974. *Linguistique générale*. Paris: Klincksieck.

Presmeg, Norma C. 1997. Reasoning with metaphors and metonymies in mathematics learning. In: English, Lyn D. (ed.), *Mathematical Reasoning: Analogies, Metaphors, and Images*. Mahwah: Lawrence Erlbaum, 267–280.

– 2005. Metaphor and metonymy in processes of semiosis in mathematics education. In: Lenhard, Johannes; Seeger, Falk (eds.), *Activity and Sign*. New York: Springer, 105–116.

Radford, Luis; Grenier, Monique 1996. On dialectical relationships between signs and algebraic ideas. *Proceedings of PME* 20: 179–186.

Radford, Luis; Schubring, Gert; Seeger, Falk (eds.) 2008. *Semiotics in Mathematics Education*. Rotterdam: Sense Publishers.

Reed, David 1994. *Figures of Thought: Mathematics and Mathematical Texts*. London: Routledge.

Richards, Ivor Armstrong 1936. *The Philosophy of Rhetoric*. Oxford: Oxford University Press.

Richeson, David S. 2008. *Euler's Gem: The Polyhedron Formula and the Birth of Topology*. Princeton: Princeton University Press.

Roberts, Royston M. 1989. *Serendipity: Accidental Discoveries in Science*. New York: John Wiley.

Rosch, Eleanor 1973a. On the internal structure of perceptual and semantic categories. In: Timothy E. Moore, (ed.), *Cognitive Development and Acquisition of Language*. New York: Academic, 111–144.

– 1973b. Natural categories. *Cognitive Psychology* 4: 328–350.

– 1975a. Cognitive reference points. *Cognitive Psychology* 7: 532–547.

– 1975b. Cognitive representations of semantic categories. *Journal of Experimental Psychology* 104: 192–233.

– 1981. Prototype classification and logical classification: The two systems. In: Scholnick, Ellyn Kofsky (ed.), *New Trends in Cognitive Representation: Challenges to Piaget's Theory*. Hillsdale: Lawrence Erlbaum Associates, 73–86.

Rosch, Eleanor; Mervis, Carolyn B. 1975. Family resemblances. *Cognitive Psychology* 7: 573–605.

Rotman, Brian 1988. Towards a semiotics of mathematics. *Semiotica* 72: 1–35.

Saichev, Alexander I.; Malevergne, Yannick; Sornette, Didier 2010. *Theory of Zipf's Law and Beyond*. New York: Springer.

Saint-Martin, Fernande 1990. *Semiotics of Visual Language*. Bloomington: Indiana University Press.

Salmon, Wesley C. 1970. *Zeno's Paradoxes*. Indianapolis: Hackett.

Saussure, Ferdinand de 1916. *Cours de linguistique générale*. [Bally, Charles; Sechehaye, Albert, eds.] Paris: Payot.

Schneider, Michael S. 1994. *Constructing the Universe: The Mathematical Archetypes of Nature, Art, and Science*. New York: Harper Collins.

Schooneveld, Cornelius H. van 1978. *Semantic Transmutations*. Bloomington: Physsardt.

Schuster, Peter 2001. *Relevance Theory Meets Markedness: Considerations on Cognitive Effort As a Criterion for Markedness in Pragmatics*. New York: Peter Lang

Sebeok, Thomas A.; Danesi, Marcel 2000. *The Forms of Meaning: Modeling Systems Theory and Semiotic Analysis*. Berlin: Mouton de Gruyter.

Semenenko, Aleksei 2012. *The Texture of Culture: An Introduction to Yuri Lotman's Semiotic Theory*. New York: Palgrave-Macmillan.

Shorser, Lindsey 2012. Manifestations of mathematical meaning. In: Bockarova, Mariana; Danesi, Marcel; Núñez, Rafael (eds.). *Semiotic and Cognitive Science Essays on the Nature of Mathematics*. Munich: Lincom Europa, 295–315.

Skemp, Richard R. 1971. *The Psychology of Learning Mathematics*. Harmondsworth: Penguin.

Smith, Edward E. 1988. Concepts and thought. In: Steinberg, Robert J.; Smith, Edward E. (eds.), *The Psychology of Human Thought*. Cambridge: Cambridge University Press, 19–49.

Stählin, Wilhelm 1914. Zür Psychologie und Statistike der Metapherm. *Archiv für Gesamte Psychologie* 31: 299–425.

Stewart, Ian 1987. *From Here to Infinity: A Guide to Today's Mathematics*. Oxford: Oxford University Press

–   1995. *Nature's Numbers*. New York: Basic Books.

–   2001. Foreword. In: Moscovich, Ivan. *1000 PlayThinks: Puzzles, Paradoxes, Illusions & Games*. New York: Workman Publishing, 1–3.

–   2012. *In Pursuit of the Unknown: 17 Equations that changed the World*. New York: Basic Books.

Stjernfelt, Frederik 2007. *Diagrammatology: An Investigation on the Borderlines of Phenomenology, Ontology, and Semiotics*. New York: Springer.

Strohmeier, John; Westbrook, Peter 1999. *Divine Harmony: The Life and Teachings of Pythagoras*. Berkeley: Berkeley Hills Books.

Swafford, Jane O.; Langrall, Cynthia W. 2000. Grade 6 students' preinstructional use of equations to describe and represent problem situations. *Journal for Research in Mathematics Education* 31: 89–110.

Taylor, John R. 1995. *Linguistic Categorization: Prototypes in Linguistic Theory*. Oxford: Oxford University Press.

Thom, René 1975. *Structural Stability and Morphogenesis: An Outline of a General Theory of Models*. Reading: Benjamin.

–   2010. Mathematics. In: Sebeok, Thomas A.; Danesi, Marcel (eds.), *Encyclopedic Dictionary of Semiotics* [3rd edition]. Berlin: Mouton de Gruyter.

Tiersma, Peter Meijes 1982. Local and general markedness. *Language* 58: 832–849.

Titchener, Edward Bradford 1910. *A Textbook of Psychology*. Delmar: Scholars' Facsimile Reprints.

Tomic, Olga Miseska (ed.) 1989. *Markedness in Synchrony and Diachrony*. Berlin: Mouton de Gruyter.

Trubetzkoy, Nikolaj S. 1936. Essaie d'une théorie des oppositions phonologiques. *Journal de Psychologie* 33, 5–18.

– 1939. Grundzüge der Phonologie. *Travaux du Cercle Linguistique de Prague* 7.

– 1968. *Introduction to the Principles of Phonological Description*. The Hague: Martinus Nijhoff.

– 1975. *Letters and Notes*. [Jakobson, Roman, ed.] The Hague: Mouton.

Turner, Mark 2012. Mental packing and unpacking in mathematics. In: Bockarova, Mariana; Danesi, Marcel; Núñez, Rafael (eds.), *Semiotic and Cognitive Science Essays on the Nature of Mathematics*. Munich: Lincom Europa, 248–294.

Twain, Mark 1889. *A Connecticut Yankee in King Arthur's Court*. New York: Dover.

Uexküll, Jakob von 1909. *Umwelt und Innenwelt der Tiere*. Berlin: Springer.

Van der Schoot, Menno; Bakker Arkema, Annemieke H.; Horsley, Tako M.; Van Lieshout, Ernest C. D. M. 2009. The consistency effect depends on markedness in less successful but not successful problem solvers: An eye movement study in primary school children. *Contemporary Educational Psychology* 34: 58–66.

Varelas, Maria 1989. Semiotic aspects of cognitive development: Illustrations from early mathematical cognition. *Psychological Review* 100: 420–431.

Vico, Giambattista 1984[1725]. *The New Science*.[2nd ed.; Bergin, Thomas Godard; Fish, Max Harold, trans.] Ithaca: Cornell University Press.

Vijayakrishnan, K. J. 2007. *The Grammar of Carnatic Music*. Berlin: Mouton de Gruyter.

Wallon, Henri 1945. *Les origines de la pensée chez l'enfant*. [Vol. 1.] Paris: Presses Universitaires de France.

Watson, Lyall 1990. *The Nature of Things*. London: Houghton and Stoughton.

Waugh, Linda 1979. Markedness and phonological systems. *LACUS (Linguistic Association of Canada and the United States) Proceedings* 5: 155–165.

– 1982. Marked and unmarked: A choice between unequals in semiotic structure. *Semiotica* 39: 211–216.

Werner, Heinz; Kaplan, Bernard 1963. *Symbol Formation: An Organismic-Developmental Approach to the Psychology of Language and the Expression of Thought*. New York: John Wiley.

Wertheimer, Max 1923. Untersuchungen zur Lehre von der Gestalt, II. *Psychologische Forschungen* 4: 301–350.

Whiteley, Walter 2012. Mathematical modeling as conceptual blending: Exploring an example within mathematics education. In: Bockarova, Mariana; Danesi, Marcel; Núñez, Rafael (eds.). *Semiotic and Cognitive Science Essays on the Nature of Mathematics*. Munich: Lincom Europa, 256–279.

Whorf, Benjamin Lee 1956. *Language, Thought, and Reality*. [Carroll, John B., ed.] Cambridge: MIT Press.

– 2012. Science and linguistics. In: Danesi, Marcel; Maida-Nicol, Sara (eds.), *Foundational Texts in Linguistic Anthropology*. Toronto: Canadian Scholars' Press, 134–156.

Wierzbicka, Anna 1996. *Semantics: Primes and Universals*. Oxford: Oxford University Press.

– 1997. *Understanding Cultures through their Key Words*. Oxford: Oxford University Press.

– 1999. *Emotions Across Languages and Cultures: Diversity and Universals*. Cambridge: Cambridge University Press.

– 2003. *Cross-Cultural Pragmatics: The Semantics of Human Interaction*. New York: Mouton de Gruyter.

Wildgen, Wolfgang; Brandt, Per Aage 2010. *Semiosis and Catastrophes: René Thom's Semiotic Heritage*. New York: Peter Lang.

Wiles, Andrew 1995. Modular elliptic curves and Fermat's last theorem. *Annals of Mathematics* 141(2nd series): 443–552.

Wiles, Andrew; Taylor, Richard 1995. Ring-theoretic properties of certain Hecke algebras. *Annals of Mathematics* 141(2nd series): 553–572.

Wundt, Wilhelm 1880. *Grundzüge der physiologischen Psychologie*. Leipzig: Engelmann.

–   1901. *Sprachgeschichte und Sprachpsychologie*. Leipzig: Eugelmann.

Zipf, George K. 1935. *Psycho-Biology of Languages*. Boston: Houghton-Mifflin.

–   1949. *Human Behavior and the Principle of Least Effort*. Boston: Addison-Wesley.

Index

Name index

Tartu Semiotics Library

Book series editors: Kalevi Kull, Silvi Salupere, Peeter Torop

Vol. 1, 1998 V. V. Ivanov, J. M. Lotman, A. M. Pjatigorski, V. N. Toporov, B. A. Uspenskij
Theses on the Semiotic Study of Cultures

Vol. 2, 1999 J. Levchenko, S. Salupere (eds.)
Conceptual Dictionary of the Tartu-Moscow Semiotic School

Vol. 3, 2002 C. Emmeche, K. Kull, F. Stjernfelt
Reading Hoffmeyer, Rethinking Biology

Vol. 4, 2005 J. Deely
Basics of Semiotics *(4th edition, bilingual)*

Vol. 4.1, 2009 J. Deely
Semiootika alused *(K. Lindström, trans.)*

Vol. 4.2, 2009 J. Deely
Basics of Semiotics *(5th edition)*

Vol. 5, 2006 M. Grishakova
The Models of Space, Time and Vision in V. Nabokov's Fiction: Narrative Strategies and Cultural Frames

Vol. 6, 2008 P. Lepik
Universaalidest Juri Lotmani semiootika kontekstis

Vol. 7, 2008 P. Lepik
Universals in the Context of Juri Lotman's Semiotics

Vol. 8, 2009 C. N. El-Hani, J. Queiroz, C. Emmeche
Genes, Information, and Semiosis

Vol. 9, 2010 A. Randviir
Ruumisemiootika: Tähendusliku maailma kaardistamine (Semiotics of Space: Mapping the Meaningful World)

Vol. 10, 2012 D. Favareau, P. Cobley, K. Kull (eds.)
A More Developed Sign: Interpreting the Work of Jesper Hoffmeyer

Vol. 11, 2012 S. Rattasepp, T. Bennett (eds.)
Gatherings in Biosemiotics

Vol. 12, 2013 F. Merrell
Meaning Making: It's What We Do; It's Who We Are — A Transdisciplinary Approach

Vol. 13, 2013 S. Salupere, P. Torop, K. Kull (eds.)
Beginnings of the Semiotics of Culture

Vol. 14, 2014 M. Danesi, M. Bockarova
Mathematics as a Modeling System

This issue has been published with
the support of the European Union through
the European Regional Development Fund
(Center of Excellence CECT)

**European Union
Regional Development Fund**   **Investing in your future**